Study Guide

Sociology
A Global Perspective

SEVENTH EDITION

Joan Ferrante

Prepared by

Joan Ferrante
Northern Kentucky University

 **WADSWORTH**
CENGAGE Learning™

Australia • Brazil • Japan • Korea • Mexico • Singapore • Spain • United Kingdom • United States

Study Guide for Sociology: A Global Perspective, Seventh Edition
Joan Ferrante

For product information and technology assistance, contact us at
Cengage Learning Customer & Sales Support,
1-800-354-9706

For permission to use material from this text or product, submit all requests online at **www.cengage.com/permissions**
Further permissions questions can be emailed to
permissionrequest@cengage.com

ISBN-13: 978-0-495-50695-9
ISBN-10: 0-495-50695-8

Wadsworth
10 Davis Drive
Belmont, CA 94002-3098
USA

Cengage Learning is a leading provider of customized learning solutions with office locations around the globe, including Singapore, the United Kingdom, Australia, Mexico, Brazil, and Japan. Locate your local office at:
www.cengage.com/international

Cengage Learning products are represented in Canada by Nelson Education, Ltd.

To learn more about Wadsworth, visit
www.cengage.com/wadsworth

Purchase any of our products at your local college store or at our preferred online store
www.ichapters.com

Printed in the United States of America
2 3 4 5 6 7 12 11 10 09 08

Contents

Preface

I wrote the *Study Guide for Sociology: A Global Perspective* 7th edition hoping to make it more than simply a test preparation tool. Everyone intuitively knows that memorizing material for the sake of doing well on a test cannot lead to meaningful and long-lasting learning experiences. In other words, if students read the text only with an eye for predicting possible test questions and take class notes without thinking about them until the night before the test, the academic experience will be an empty one. Real learning means thinking about the ideas that you hear and read, taking an active role in learning, and incorporating learning experiences into the activities of daily life. I wrote the Study Guide with this vision of learning in mind. Each chapter corresponds to a chapter in the textbook and includes seven sections: (1) study questions, (2) concept application, (3) practice multiple-choice and true/false questions, (4) applied research, (5) internet resources, and (6) statistical profiles. In addition, please visit the supporting Web site at http://sociology.wadsworth.com. Just click on the book cover icon under the Introductory Sociology heading.

Study Questions

Think of the study questions as note-taking tools. If you answer these questions thoughtfully and conscientiously, you will come away with a thorough review of the sociological material covered in the textbook.

Concept Applications

Each chapter contains several scenarios, about a paragraph long to read and then to decide which sociological concept best fits the scenario. You are asked to explain why.

Practice Multiple-Choice and True/False Questions

If you treat these questions as a comprehensive study tool, you will not be adequately prepared to take a test. On the other hand, if you prepare beforehand (as if you are going to take a test that will count for a grade), the score you earn will probably be a good indicator of how well you can expect to do on the actual test.

Applied Research

The applied research exercises take learning beyond the book and ask students to become learners that are actively involved in gathering information to answer questions that relate to material covered in the textbook.

Internet Resources

For each chapter there are recommended internet resources related to the chapter topic and to the country emphasized in that chapter.

Country Statistical Profile

Eleven of the sixteen chapters in *Sociology: A Global Perspective* incorporate information about life in countries other than the United States. Usually just reading the chapter gives sufficient background information on the countries emphasized. For those who would like additional coverage I have included additional background material of a general nature.

Chapter 1

The Sociological Imagination

Study Questions

1. What is sociology? What do sociologists study?

2. What is social interaction?

3. Durkheim maintains that the sociologist's task is to study social facts. What are social facts?

4. When do people experience the power of social facts?

5. In the classic book *Invitation to Sociology*, Peter L. Berger presents sociology as a form of consciousness. Explain.

6. In studying patterns of courtship and marriage, what would sociologists emphasize?

7. Peter Berger maintains that a "debunking motif" defines the sociological consciousness. Explain.

8. Give two examples of the kinds of questions sociologists ask.

9. Distinguish between troubles and issues.

10. Explain the connection between troubles and institutional crisis.

11. What is the sociological imagination?

12. What major historical event shaped the discipline of sociology? Why?

13. How did the Industrial Revolution affect the nature of work and social interaction?

14. Who was Auguste Comte? How did he define sociology? Does his definition speak to the importance of social interaction?

15. For which writings is Marx most famous?

16. Define Karl Marx's vision of the sociologist's task. What concepts and assumptions drive his analysis of society?

17. Who are the bourgeoisie and the proletariat? How are they connected to the means of production?

18. How did Durkheim define suicide?

19. Distinguish between egoistic, altruistic, anomic, and fatalistic suicide.

20. What is social action? What are the four types? Give an example of each.

21. Who is Harriet Martineau? What contributions did she make to sociology?

22. Explain the phrase "strange meaning of being black." What life experience may have influenced DuBois' preoccupation with this phrase?

23. According to DuBois, how did the color line come into being?

24. What is double consciousness?

25. Describe three assumptions that underlie the global perspective.

26. Imagine that you majored in sociology. How would you explain the usefulness of the sociological perspective? What skills would you bring to the workplace?

Concept Applications

Consider the concepts listed below. Match one or more of the concepts with each scenario. Explain your choices.

a. Anomic
b. Double consciousness
c. Global interdependence
d. Means of production
e. Troubles/Issues

Scenario 1

Excerpts from a suicide letter suggest that Kevin Morrissey, a 51-year-old Berkeley man, killed his family in a murder-suicide this week because he was at a "financial breaking point" as the family skin-care business failed and because he found other work opportunities "unattractive" (Rayburn and Hill 2007).

Scenario 2

"The world's trade in bananas is dominated by just three huge food multinationals: United Brands (with a 34 percent market share in 1974), Standard Fruit (with a 23 percent market share), and Del Monte (with a 10 percent market share). As with many other commodities, the companies control the transport, packaging, shipment, storage, and marketing of the fruit. As a result, the profits from bananas go largely into western pockets, while the producer countries get only a pittance" (Harrison 1987:348).

Scenario 3

"*Black Soldiers in Jim Crow Texas* introduces readers to African American soldiers who were assigned to one of four black regiments (9th and 10th Cavalries and 24th and 25th Infantries). Not only did these men bear arms and fight gallantly in the Spanish-American War, but at times, they used their military weapons in struggles for racial equality in the United States as well. More than three decades after the Emancipation Proclamation, black soldiers grew intolerant of 'racial slurs, refusal of service at some businesses, and harassment.' Texas's 'lower-status Hispanics, the bulk of the population…shared southern white prejudice against blacks. The war with Spain in 1898,' Christian asserts, 'acted as a catalyst that converted impatience into retaliation. The United States bestowed six Medals of Honor and twenty-six Certificates of Merit on their members, and all four regiments inspired laudatory press coverage.' Yet these men faced the indignities of racism when serving at military installations in the United States" (Moore 1996:478).

Scenario 4 (new)

On the progressive care unit where she works, nurses regularly have five or more patients. Over the years, hospital procedures with which nurses assist have become more complicated, and patients are sicker. Brandon said there are not always enough nurses to go around. "You get your running shoes on, take off, and go," Brandon said. "The current nursing shortage is just beginning in Wyoming," said Julie Cann-Taylor, registered nurse and director of critical care at the hospital. "There had been a nursing vacancy rate of 3 to 4 percent at the hospital for years, but it jumped to 7 percent last fall," she said. Matt Kaiser, director of human resources at the hospital, said there are about 40 registered nurse positions available, creating a vacancy rate of about 11 percent (Rupp 2007).

Scenario 5

Five foreign-born players appeared in the National Basketball Association All-Star game last month, and another five played in the Rookie Challenge game. Of the 348 active players in the NBA, 49 are from abroad…Lenny Wilken, coach of the Toronto Rapters, has said, "I wouldn't be surprised if there is double the number of players in the next five years or so." (Shield's 2002:56)

Practice Test: Multiple-Choice Questions

1. Sociologists maintain that there is a system to interaction. "System" means that
   a. interaction is chaotic.
   b. established rules (often unspoken) guide interaction.
   c. interaction is unpredictable.
   d. people make up the rules as they interact.

2. When sociologists study social interaction, they focus on all *but* which one of the following?
   a. the ways in which people who do *not* know each other manage to interact
   b. the system guiding social interaction
   c. the personalities of those involved
   d. the spoken and unspoken rules guiding that interaction

3. A woman writes, "I can't be anything but what my skin color tells people I am. I am black because I look black. It does not matter that my family has a complicated biological heritage." She is writing about the power of
   a. social facts.
   b. troubles.
   c. the sociological imagination.
   d. rationalization.

4. In the United States, approximately _____ percent of people have been married at least once by age 75.
   a. 95
   b. 75
   c. 50
   d. 33

5. A sociologist views a photo of an American soldier and an Iraqi child bumping fists. The image prompts the sociologist to ask:
   a. What does it mean for the U.S. to occupy/liberate a country where 40 percent of the population is 14 and under?
   b. Does the American soldier have a child of his own?
   c. Is the soldier an occupier or a liberator?
   d. How many American soldiers are stationed in Iraq?

6. Which of the following explanations would someone use to explain an issue?
    a. "She had the opportunity but didn't take it."
    b. "He is lazy."
    c. "There is a flaw or breakdown in an institutional arrangement."
    d. "She didn't try very hard in school."

7. The high school dropout rate in the United States is greater than 25 percent. C. Wright Mills would classify this situation as
    a. a trouble.
    b. an issue.
    c. value-rational action.
    d. a social fact.

8. The sociological imagination is the ability to
    a. see the connection between self and immediate relationships.
    b. distinguish between mechanical and organic solidarity.
    c. see that problems can be solved by changing the character of the individual.
    d. make a distinction between troubles and issues.

9. The Industrial Revolution transformed the nature of work in which one of the following ways?
    a. Machine production was replaced by hand production.
    b. People now could say, "I made this; this is a unique product of my labor."
    c. Products became standardized, and workers performed specific tasks in the production process.
    d. The workers' power over the production process increased dramatically.

10. The name *sociology* and the corresponding academic discipline was born during the
    a. American Revolution.
    b. Civil War.
    c. Vietnam era.
    d. Industrial Revolution.

11. Marx's legacy has been obscured by
    a. his inability to accurately describe capitalism.
    b. a personality disorder.
    c. the failure of Communism.
    d. the fact that he published in German (not English).

12. Land, tools, equipment, factories, modes of transportation, and labor are
    a. owned by the proletariat.
    b. part of the means of production.
    c. essential for providing services.
    d. owned by the intellectual classes.

13. From a sociological perspective, suicide is
    a. an act of intentionally killing oneself.
    b. the result of personal disappointment and sorrow.
    c. self-hatred actualized.
    d. the severing of relationships.

14. When people commit _____ suicide, it is on behalf of the group they love more than themselves.
    a. egoistic
    b. altruistic
    c. anomic
    d. fatalistic

15. If an individual pursues a college degree because everyone in his or her family going back five generations is college-educated, the action can be classified as
    a. traditional.
    b. affectional.
    c. value-rational.
    d. instrumental.

16. W.E.B. DuBois coined the phrase
    a. the ties that bind people to one another.
    b. the "strange meaning of being black."
    c. the means of production.
    d. the course and consequences of social action.

For the following questions, use one of these responses to identify the thinker associated with each statement.

    a. Karl Marx
    b. Emile Durkheim
    c. Max Weber
    d. Harriet Martineau
    e. W.E.B. DuBois

17. The sociologist's task is to study social facts.

18. In conducting social research, it is important to see a country in all its diversity.

19. Capitalism has unleashed "wonders far surpassing Egyptian pyramids, Roman aqueducts, and Gothic cathedrals."

20. Which one of the following statements would be most likely to convince an employer of the worth of a sociology degree?
    a. "I like people, and sociology is about people."
    b. "I want to work with people. That is why I majored in sociology."
    c. "I didn't have to take a statistics course."
    d. "Among other things, a degree in sociology helps me to identify and project population trends."

True/False Questions

1. T F The interactions sociologists study can involve two people or thousands of people.

2. T F Sociologists focus on the role personality plays in driving social interactions.

3. T F Sociologists maintain that love is a violent, irresistible emotion that strikes someone at random.

4. T F Nineteen of 20 people get married at least once in their lifetime.

5. T F Sociologist C. Wright Mills argues that most people cannot or do not want to see how their successes connect to others' failures.

6. T F The changes triggered by the Industrial Revolution are incalculable.

7. T F In analyzing suicide rates, Durkheim emphasized the personal situation of the victim.

8. T F DuBois believed that the problem of the twentieth century was the problem of the color line.

9. T F Globalization is a relatively new phenomenon, which can be traced to the 1990s.

10. T F A degree in sociology leads to very few career tracks.

Internet Resources

- **Sociological Tour Through Cyberspace**
  http://www.trinity.edu/~mkearl/
  Sociologist Michael Kearl at Trinity University is interested in cyberspace's potential "to inform and generate discourse, to truly be a 'theater of ideas'". To demonstrate this potential Kearl has created more than 20 such "theaters," which explore topics of interest to any student of sociology, including marriage and family life, social gerontology, social inequality, gender and society, race and ethnicity, and sociology of death and dying.

- **YaleGlobal On-Line**
  http://yaleglobal.yale.edu/globalization/
  "Debate abounds over whether globalization is good or bad for the self, the family, the nation, and the world. Some pessimists see increased interdependence as a terribly destructive trend, while optimists see a more diverse, better life for all. Some people argue that the world is no more globalized than it was in the waning days of the British Empire, but some see an information revolution that is unparalleled in history and widespread in its implications." Yale Global On-Line has posts a assembled a series of articles to shed light on this debate."

Applied Research

Find a newspaper or magazine article in which the reporter highlights a seemingly personal problem. Briefly describe the problem. Does the article suggest a cause of the problem? Does the article connect the individual trouble to a larger issue or to flaws or breakdowns in institutional arrangements? If yes, explain. If no, can the problem be explained in terms of a larger issue? How so?

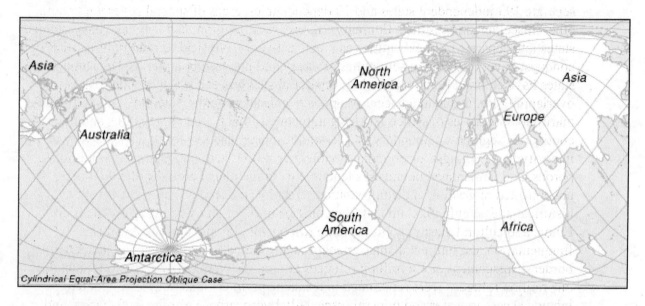

Population: (2007) 6.6 billion
Infant mortality rate: 43.52 deaths/1,000 live births
Religions: Christians 33.03% (of which Roman Catholics 17.33%, Protestants 5.8%, Orthodox 3.42%, Anglicans 1.23%), Muslims 20.12%, Hindus 13.34%, Buddhists 5.89%, Sikhs 0.39%, Jews 0.23%, other religions 12.61%, non-religious 12.03%, atheists 2.36% (2004 est.)
First Languages: Mandarin Chinese 13.69%, Spanish 5.05%, English 4.84%, Hindi 2.82%, Portuguese 2.77%, Bengali 2.68%, Russian 2.27%, Japanese 1.99%, Standard German 1.49%, Wu Chinese 1.21% (2004 est.)

Globally, the 20th century was marked by: (a) two devastating world wars; (b) the Great Depression of the 1930s; (c) the end of vast colonial empires; (d) rapid advances in science and technology, from the first airplane flight at Kitty Hawk, North Carolina (US), to the landing on the moon; (e) the Cold War between the Western alliance and the Warsaw Pact nations; (f) a sharp rise in living standards in North America, Europe, and Japan; (g) increased concerns about the environment, including loss of forests, shortages of energy and water, the decline in biological diversity, and air pollution; (h) the onset of the AIDS epidemic; and (i) the ultimate emergence of the US as the only world superpower. The planet's population continues to explode (from 1 billion in 1820 to 2 billion in 1930 to 3 billion in 1960 to 4 billion in 1974 to 5 billion in 1988 to 6 billion in 2000). For the 21st century, the continued exponential growth in science and technology raises both hopes (e.g., advances in medicine) and fears (e.g., development of even more lethal weapons of war).

Stretching over 250,000 kilometers, the world's 319 international land boundaries separate 193 independent states and 70 dependencies, areas of special sovereignty, and other miscellaneous entities. Ethnicity, culture, race, religion, and language have divided states into separate political entities as much as history, physical terrain, political fiat, or conquest have, resulting in sometimes arbitrary and imposed boundaries. Most maritime states have claimed limits that include territorial seas and exclusive economic zones; overlapping limits due to adjacent or opposite coasts create the potential for 430 bilateral maritime boundaries of which 209 have agreements that include contiguous and non-contiguous segments. Boundary, borderland/resource, and territorial disputes vary in intensity from managed or dormant to violent or militarized; undemarcated, indefinite, porous, and unmanaged boundaries tend to encourage illegal cross-border activities, uncontrolled migration, and confrontation. Territorial disputes may evolve from historical and/or cultural claims, or they may be brought on by resource competition. Ethnic and cultural clashes continue to be responsible for much of the territorial fragmentation and internal displacement of an estimated 6.6 million people and cross-border displacement of 8.6 million refugees around the world as of early 2006; just over one million refugees were repatriated in the same period. Other sources of contention include access to water and mineral (especially hydrocarbon) resources, fisheries, and arable land. Armed conflict prevails not so much between the uniformed armed forces of independent states as between stateless armed entities that detract from the sustenance and welfare of local populations, leaving the community of nations to cope with resultant refugees, hunger, disease, impoverishment, and environmental degradation

Global output rose by 5% in 2006 and was led by China (10.5%), India (8.5%), and Russia (6.6%). The 14 other successor nations of the USSR and the other old Warsaw Pact nations again experienced widely divergent growth rates; the three Baltic nations continued as strong performers, in the seven to ten percent range of growth. Growth results posted by the major industrial countries varied from no gain for Italy to a strong gain by the United States (3.4%). The developing nations also varied in their growth results, with many countries facing population increases that eroded gains in output. Externally, the nation-state, as a bedrock economic-political institution, is steadily losing control over international flows of people, goods, funds, and technology. Internally, the central government often finds its control over resources slipping as separatist regional movements—typically based on ethnicity—gain momentum (e.g., in many of the successor states of the former Soviet Union, in former Yugoslavia, in India, in Iraq, in Indonesia, and in Canada). Externally, the central government is losing decision making powers to international bodies, notably the EU. In Western Europe, governments face the difficult political problem of channeling resources away from welfare programs in order to increase investment and strengthen incentives for individuals to seek employment. The addition of 80 million people each year to an already overcrowded globe is exacerbating the problems of pollution, desertification, underemployment, epidemics, and famine. Because of their own internal problems and priorities, the industrialized countries devote insufficient resources to deal effectively with the poorer areas of the world, which are becoming further marginalized, at least from an economic point of view. The introduction of the euro as the common currency of much of Western Europe in January 1999, while paving the way for an integrated economic powerhouse, posed economic risks because of varying levels of income and cultural and political differences among the participating nations. The terrorist attacks on the US on September 11, 2001, accentuated a further growing risk to global prosperity, as is illustrated by the example of the reallocation of resources away from investment programs and toward anti-terrorist programs. In March 2003, the opening of war between a US-led coalition and Iraq added new uncertainties to global economic prospects. After the coalition victory, the complex political difficulties and the high economic costs of establishing domestic order in Iraq became major global problems that continued through 2006.

Source: Excerpted from U.S. Central Intelligence, 2007 World Factbook
www.cia.gov/library/publications/the-world-factbook/geos/xx.ht

Answers

Concept Applications
1. Anomic
2. Means of production
3. Double consciousness
4. Trouble/Issues
5. Global interdependence

Multiple-Choice

1.	b	page 1	11.	c	page 13
2.	c	page 5	12.	b	page 14
3.	a	page 6	13.	d	page 15
4.	a	page 6	14.	b	page 15
5.	a	page 8	15.	a	page 18
6.	c	page 11	16.	b	page 18
7.	b	page 10	17.	b	page 5
8.	d	page 11	18.	d	page 18
9.	c	page 12	19.	a	page 13
10.	d	page 13	20.	d	page 22

True/False

1.	T	page 2
2.	F	page 5
3.	F	page 6
4.	T	page 6
5.	T	page 10
6.	T	page 12
7.	F	page 15
8.	T	page 18
9.	F	page 20
10.	F	page 20

Chapter 2

Theoretical Perspectives and Methods of Social Research

Study Questions

1. Why is Mexico (in particular, the border fence) the focus of Chapter 2?

2. What are the names of the three theoretical perspectives? How do the three theoretical perspectives help us to think about any social event or issue?

3. What is a function? Give an example.

4. According to the functionalist perspective, why has poverty not been eliminated?

5. What concepts did Robert K. Merton introduce to counter criticisms of the functionalist perspective? Briefly define each concept, and explain how they strengthen the perspective. What criticism is not addressed by Merton's concepts?

6. Use the following chart to summarize a functionalist analysis of community-wide celebrations.

	Function	Dysfunction
Manifest		
Latent		

7. For what reasons did the United States construct fences (and plan to construct more fences) along the U.S.-Mexico border?

8. List one example of a manifest function, manifest dysfunction, latent function, and latent dysfunction associated with the construction of border fences.

9. What question guides conflict theorists is their analysis of any social issue? In answering that question, what do conflict theorists emphasize?

10. How does a conflict theorist explain the purpose of the fences along the U.S.-Mexico border?

11. What are international remittances? Explain their importance.

12. What central concepts and questions guide the symbolic interactionist perspective?

13. How would symbolic interactionists study border crossing?

14. What are the strengths and weaknesses of each theoretical perspective?

15. Define research methods.

16. What assumptions underlie the scientific method? Under what circumstances do research findings endure? Contrast the ideal of the research process with reality.

17. Why is it important for researchers to explain their reasons for choosing to investigate a particular topic? Why did Singer and Massey study undocumented border crossings?

18. Why should researchers review the literature before beginning to investigate a topic?

19. What are concepts, and how do they relate to the research process?

20. What kinds of "things" do sociologists study? Give examples.

21. Why do sociologists study random samples? Why are random samples difficult to secure? Under what conditions are nonrandom samples acceptable?

22. Give a brief description of each method of data collection.

    Self-Administered Questionnaires

    Structured Interviews

    Unstructured Interviews

    Participant Observation

    Secondary Sources

23. What is the Hawthorne effect?

24. What is a hypothesis? Give an example of a hypothesis. Identify the independent and dependent variables.

25. What is an operational definition? Give an example.

26. Distinguish between reliability and validity.

27. How do basic statistics help to describe apprehensions involving illegal immigrants over the past 12 years?

28. What is generalizability? Under what conditions can findings be considered generalizable?

29. What three conditions must be met before a researcher can claim that an independent variable contributes significantly to explaining a dependent variable?

Concept Applications

Consider the concepts listed below. Match one or more of the concepts with each scenario. Explain your choices.

a. Function
b. Dysfunction
c. Symbol
d. Territories
e. Participant observation
f. Spurious correlation
g. Proletariat

Scenario 1

"The influx of Korean-owned firms conferred obvious economic benefits on Los Angeles. (1) Korean firms tended to service low income, nonwhite neighborhoods generally ignored and underserved by big corporations….(2) The Korean influx restored the [deteriorating and underutilized] neighborhoods in which Koreans settled….(3) Their residential and commercial interests compelled Koreans to combat street crime….(4) Koreans valued public education and improved it. Indeed, many Korean families had emigrated to the United States because of this country's superior educational opportunities" (Light and Bonacich 1988:6-7).

Scenario 2

"Dr. Louise Keating became 'Trash Czar' for a few days. Dr. Keating, director of Red Cross Blood Services in Cleveland, found her center almost engulfed by mounds of debris—dressings, needles, plastic tubes—most of it the usual detritus of any organization, but some of it splashed with the blood of donors. Her center was not generating any more trash than usual. But suddenly, no one was willing to cart it away. AIDS could be transmitted through blood, we had now learned. Last year's innocuous garbage had become this year's plague vector. Or so it seemed to Cleveland's carters. And the refuse piles grew."

"Dr. Keating did solve her problem. Now, all waste that has any blood on it is sterilized in an autoclave until nothing, not even a virus, survives. But AIDS has created many other problems in the nation's blood supply: for those, like Dr. Keating and her colleagues, who must find donors and ensure that the blood obtained is safe; for those who give blood; and for those who receive it" (Murray 1990:205).

Scenario 3

Some Americans venturing into Mexico probably hear the word [gringo] and wonder if somebody is picking a fight. The answer seems to depend on who says it and how. "It's all in the tone; usually the eyes will tell you something as well," said Tony Garza, the U.S. ambassador to Mexico, who grew up in Brownsville. "It can mean everything from 'I am going to try and kick your butt,' to 'friend, let's have a drink,'" Garza added. "Let's jus say it is very situation-specific." When gringo is used in Mexico, it tends to be applied to anyone born in the United States, regardless of race or background (Schiller 2004).

Scenario 4

"I gained access to the enterprise through a friend who was a manager in a local bank from which the enterprise borrowed commercial loans. Management and workers in both factories knew I was a graduate student writing a dissertation. I was a full-time assembly worker in the Hong Kong plant, visited workers' homes, and participated in their weekend activities. In Shenzhen, I observed and talked with workers and managers on the shop floor and the office, but management allowed me to work on the line only occasionally. I lived in factory dormitories together with other Hong Kong managerial staff, but I visited and interviewed workers in workers' dormitories. I also participated in both workers' and managers' gatherings after work" (Lee 1995:380).

Scenario 5

For the class, the suburban mall became the microsocial setting for investigating macrotheoretical issues. Students examined specific features of their selected malls, such as the surrounding physical environment (entrance, parking, sidewalks), financial condition (unoccupied spaces, needed repairs, open-air merchants), design of interior space (escalators, lighting, plants), types of stores (prestige, anchors, discounters, specialties), clientele (social class, gender, race, ethnicity), nationality as well as race and ethnicity and gender of merchants (especially subcontractors within stores) and employees, pricing structure (including types of credit cards accepted or interest-free purchase options), mall names and distinctive linguistic terms, treatment of shoppers by employees, safety and security issues, and the presence of "mall zombies" as a crude indicator of the dehumanizing effects associated with "irrationality of rationality" (Manning, Price, and Rich 1997:18).

Scenario 6

There is a positive correlation between ice cream sales and deaths due to drowning: the more ice cream sold, the more drownings and vice versa. The third variable at work here is *season* or *temperature*. Most drowning deaths occur during the warm days of summer—and that's the peak period for ice cream sales. There is no direct link between ice cream and drowning (Babbie 1995:70).

Scenario 7

In April, KenSa started production in a rented, temporary facility in San Pedro Sula while a new factory was being built. It hired 150 workers to make Chrysler minivan door wire harnesses at the Honduran minimum wage of about 55 cents an hour. It turned hundreds of people away…But the jobs provide no economic miracle. Factory work barely provides enough to live…Most workers at foreign-owned factories in Honduras make $4.44 a day. That's less than the $5 a day Henry Ford paid his Highland Park workers 90 years ago in 1914. Ford's pay (the equivalent of $11 an hour today) more than doubled the minimum wage at the time and helped give birth to America's blue-collar middle class…U.S. companies have no such incentive in countries such as Honduras. Products are built for export back to America. Raising worker salaries in San Pedro Sula won't sell even one more SUV in Detroit (French 2004).

Practice Test: Multiple-Choice Questions

1. The United States shares a _____-mile long border with Mexico.
    a. 200
    b. 800
    c. 2,000
    d. 5,000

2. According to functionalists, poverty exists because
    a. the poor lack skills to do better.
    b. it contributes in some way to the stability of the overall society.
    c. the poor lack the drive to do better.
    d. somebody has to be on the bottom.

3. Sometimes police departments choose to negotiate contracts with the host city just before a community-wide celebration, thereby using the event as a bargaining tool to secure a good contract. From the *city's perspective*, this represents a _____ of the community-wide celebration.
   a. latent dysfunction
   b. latent function
   c. manifest function
   d. facade of legitimacy

4. An *unexpected outcome* of the border fence is the emergence of humanitarian groups that save the lives of many illegal immigrants but, in doing so, help people circumvent the law. This outcome is an example of a
   a. manifest function.
   b. manifest dysfunction.
   c. latent function.
   d. latent dysfunction.

5. Conflict theorists are inspired by
   a. Max Weber.
   b. Emile Durkheim.
   c. Karl Marx.
   d. C. Wright Mills.

6. The worker "has but wages to live upon, and must therefore take work when, where, and at what terms he can get it." This line, written in 1881, applies to the situation of the
   a. proletariat.
   b. bourgeoisie.
   c. means of production.
   d. capitalist.

7. "Meaning is not evident from the physical phenomenon alone." This statement suggests that
   a. people assign meaning to that phenomenon.
   b. meaning is fixed and universal.
   c. people can tell what something "means" just by looking at it.
   d. seeing is believing.

8. Which one of the following statements represents a criticism of the functionalist perspective?
   a. It is too liberal.
   b. It focuses on the "small stuff."
   c. It offers no technique for determining the "overall net effect."
   d. It focuses on the "have nots."

9. Which one of the following questions about illegal immigration from Mexico to the U.S. would be of most interest to a conflict theorist?
   a. How does illegal immigration contribute to order and stability in Mexico and the U.S.?
   b. Who benefits from the existence of illegal immigration, and at whose expense?
   c. Does everyone in the U.S. and Mexico see illegal immigration in the same way?
   d. Why doesn't the U.S close its borders to foreign workers?

10. Which one of the following assumptions applies to the scientific method?
    a. Knowledge is always subjective.
    b. Research findings can be manipulated to advance a good cause.
    c. Truth is confirmed through faith.
    d. Knowledge is acquired through observation.

11. Sociological research is guided by
    a. methods unique to the discipline.
    b. a passion to change society.
    c. emotion and personal interest.
    d. the scientific method.

12. In theory, the first step in undertaking a sociological research project is
    a. consulting existing research.
    b. collecting data.
    c. choosing a topic for investigation.
    d. analyzing the data.

13. Sociologists Audrey Singer and Douglas S. Massey maintain that "constructing fences and implementing other border control strategies sits well with the public as the government appears to be defending the United States against alien invaders while not antagonizing U.S. business interests." This statement suggests the two sociologists are taking a _____ perspective to frame their research.
    a. functionalist
    b. conflict
    c. symbolic interaction
    d. sociological

14. _____ are materials or other evidence that yields information about human activity, including items that people throw away or the number of lights left on in homes at a particular time.
    a. Traces
    b. Documents
    c. Territories
    d. Households

15. For his book *Patrolling Chaos*, sociologist Robert Maril accompanied 12 border patrol agents on 60 ten-hour shifts along the border. Maril had chosen to study
    a. small groups.
    b. documents.
    c. territories.
    d. households.

16. _____ is especially useful for studying behavior as it occurs.
    a. A self-administered questionnaire
    b. Secondary data analysis
    c. An interview
    d. Observation

17. One of the primary reasons researchers engaged in participant observation conceal their identity is to eliminate
    a. legal problems.
    b. the need for confidentiality.
    c. the Hawthorne effect.
    d. ethical considerations.

18. In research, the variable to be explained or predicted is known as
    a. the dependent variable.
    b. the independent variable.
    c. the hypothesis.
    d. the control variable.

19. The *dependent* variable in the hypothesis "the longer a U.S. line worker has been employed at a U.S.-based assembly plant, the more time it takes for that worker to find new employment when the assembly plant moves to Mexico" is
    a. employment at U.S. based assembly plant.
    b. assembly plants in Mexico.
    c. the length of time employed at line work in U.S. assembly plant.
    d. the length of time to find new employment.

20. A professor tells a class that exams will cover information from class lectures, class discussion, and reading assignments. However, the exam includes questions related to only reading assignments. Students complain because the exam is
    a. not reliable.
    b. not valid.
    c. not reliable or valid.
    d. objective.

True/False Questions

1. T F From a functionalist viewpoint, poverty contributes to the stability of the overall society.

2. T F "Latent" means intended, anticipated, or expected.

3. T F The border fences have forced illegal immigrants to enter the United States through desert and other inhospitable terrain.

4. T F The façade of legitimacy is an explanation that members of dominant groups give to justify exploitive actions.

5. T F Many households in Mexico have come to rely on remittance income.

6. T F Symbolic interactionists focus on social interaction.

7. T F One strength of the symbolic interactionist perspective is that it gives a balanced overview of intended and unintended consequences.

8. T F Researchers do not always follow in order the steps of scientific method.

9. T F Structured interviews are flexible and open-ended in style.

10. T F A correlation of -1.0 suggests that there is no relationship between two variables.

Internet Resources

Sociological Tour Through Cyberspace
http://www.trinity.edu/~mkearl/
Sociologist Michael Kearl at Trinity University is interested in cyberspace's potential "to inform and generate discourse, to truly be a 'theater of ideas.'" To demonstrate this potential Kearl has created   more than 20 such "theaters," which explore topics of interest to any student of sociology, including marriage and family life, social gerontology, social inequality, gender and society, race and ethnicity, and sociology of death and dying.

Applied Research

Find a research article in a sociological journal such as *Sociological Focus, Journal of Comparative Sociology,* or *American Sociological Review*. Give a brief description of the following elements:

Topic
Major concepts or theoretical perspectives
Independent and dependent variables (if applicable)
Unit of analysis
Methods of data collection
Operational definition (if applicable)
Findings
Conclusion

YaleGlobal On-Line
http://yaleglobal.yale.edu/globalization/
"Debate abounds over whether globalization is good or bad for the self, the family, the nation, and the world. Some pessimists see increased interdependence as a terribly destructive trend, while optimists see a more diverse, better life for all. Some people argue that the world is no more globalized than it was in the waning days of the British Empire, but some see an information revolution that is unparalleled in history and widespread in its implications." Yale Global On-Line has posts a assembled a series of articles to shed light on this debate."

Background Notes: United Mexican States (Mexico)

Size: About three times the size of Texas.
Capital: Mexico City (18.7 million)
Population (2006): 107.4 million.
Ethnic groups: Indian-Spanish (mestizo) 60%, Indian 30%, Caucasian 9%, other 1%.
Religions: Roman Catholic 89%, Protestant 6%, other 5%.
Infant mortality rate: 21.69/1000.
Life expectancy: male 72.18 years; female 77.83 years.
Work force (2005): 39.81 million:

Workforce by Sector: Agriculture, forestry, hunting, fishing--21.0%; services--32.2%; commerce--16.9%; manufacturing--18.7%; construction--5.6%; transportation and communication--4.5%; mining and quarrying--1.0%.

Mexico is the most populous Spanish-speaking country in the world and the second most-populous country in Latin America after Portuguese-speaking Brazil. About 70% of the people live in urban areas. Many Mexicans emigrate from rural areas that lack job opportunities--such as the underdeveloped southern states and the crowded central plateau--to the industrialized urban centers and the developing areas along the U.S.-Mexico border. According to some estimates, the population of the area around Mexico City is about 18 million, which would make it the largest concentration of population in the Western Hemisphere. Cities bordering on the United States--such as Tijuana and Ciudad Juarez--and cities in the interior--such as Guadalajara, Monterrey, and Puebla--have undergone sharp rises in population in recent years.

Mexico is highly dependent on exports to the U.S., which account for almost a quarter of the country's GDP. The result is that the Mexican economy is strongly linked to the U.S. business cycle. Real GDP grew by 3.0% in 2005 and was projected to grow by 4.5% for 2006. Mexico's trade regime is among the most open in the world, with free trade agreements with the U.S., Canada, the EU, and many other countries. Since the 1994 devaluation of the peso, successive Mexican governments have improved the country's macroeconomic fundamentals.

Mexico is among the world's most open economies, but it is dependent on trade with the U.S., which bought 86% of its exports in 2005. Top U.S. exports to Mexico include electronic equipment, motor vehicle parts, and chemicals. Top Mexican exports to the U.S. include petroleum, cars, and electronic equipment. There is considerable intra-company trade. . . The most significant areas of friction involve agricultural products such as sugar, high fructose corn syrup, apples, and rice.

In 2005 Mexico was the world's sixth-largest oil producer, its eighth-largest oil exporter, and the third-largest supplier of oil to the U.S. Oil and gas revenues provide more than one-third of all Mexican Government revenues. Mexico's state-owned oil company, Pemex, holds a constitutionally established monopoly for the exploration, production, transportation, and marketing of the nation's oil. While private investment in natural gas transportation, distribution, and storage has been permitted, Pemex remains in sole control of natural gas exploration and production. Despite substantial reserves, Mexico is a net natural gas importer.

U.S. relations with Mexico are as important and complex as with any country in the world. A stable, democratic, and economically prosperous Mexico is fundamental to U.S. interests. U.S. relations with Mexico have a direct impact on the lives and livelihoods of millions of Americans--whether the issue is trade and economic reform, homeland security, drug control, migration, or the promotion of democracy. The U.S. and Mexico are partners in NAFTA, and enjoy a rapidly developing trade relationship. . . The scope of U.S.-Mexican relations goes far beyond diplomatic and official contacts; it entails extensive commercial, cultural, and educational ties, as demonstrated by the annual figure of nearly a million legal border crossings a day. In addition, more than a half-million American citizens live in Mexico. More than 2,600 U.S. companies have operations there, and the U.S. accounts for 55% of all foreign direct investment in Mexico. Along the 2,000-mile shared border, state and local governments interact closely.

Cooperation between the United States and Mexico along the 2,000-mile common border includes state and local problem-solving mechanisms; transportation planning; and institutions to address resource, environment, and health issues…As the number of people and the volume of cargo crossing the U.S.-Mexico border grow, so, too, does the need for coordinated infrastructure development. The multi-agency U.S.-Mexico Binational Group on Bridges and Border Crossings meets twice yearly to improve the efficiency of existing crossings and coordinate planning for new ones. The 10 U.S. and Mexican border states have become active participants in these meetings.

Source: Excepted from U.S. Department of State: Background Notes
http://www.state.gov/r/pa/ei/bgn/35749.htm

Answers

Concept Applications
1. Function
2. Dysfunction
3. Symbol
4. Participant observation
5. Territories
6. Spurious correlation
7. Proletariat

Multiple Choice

1.	c	page 27	11.	d	page 40
2.	b	page 29	12.	c	page 41
3.	a	page 30	13.	b	page 41
4.	d	page 31	14.	a	page 42
5.	c	page 32	15.	a	page 44
6.	a	page 33	16.	d	page 45
7.	a	page 37	17.	c	page 47
8.	c	page 39	18.	a	page 47
9.	b	page 39	19.	d	page 47
10.	d	page 40	20.	b	page 48

True/False

1.	T	page 29
2.	F	page 29
3.	T	page 39
4.	T	page 32
5.	T	page 35
6.	T	page 35
7.	F	page 39
8.	T	page 41
9.	F	page 45
10.	F	page 51

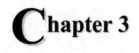hapter 3

Culture

Study Questions

1. Why were North and South Korea chosen to illustrate the concept of culture?

2. How is the word "culture" typically used among English speakers?

3. What three conceptual challenges do sociologists face in defining culture?

4. Distinguish between nonmaterial and material culture.

5. Distinguish between beliefs and values. Give an example of each.

6.   What are norms? Distinguish between folkways and mores.

7.   What are symbols? Is language a symbol? Explain.

8.   How do geographic and historical forces shape culture?

9.   Explain: "Regardless of their physical traits, babies are destined to learn the ways of the culture into which they are born and raised."

10. In what ways does language channel thinking?

11. What is the linguistic relativity hypothesis?

12. How are people products of cultural experiences, yet not cultural replicas of one another?

13. Explain: "All cultures have developed formulas to help their members respond to biological inevitabilities."

14. What makes an emotion social? What is the connection between feeling rules and social emotions?

15. What is diffusion? Give two examples of the diffusion process. Why is diffusion a selective process?

16. Give two examples of opportunities for cultural diffusion between Americans and South Koreans.

17. What is culture shock? How is it related to ethnocentrism?

18. What are the various types of ethnocentrism? Give examples of each. (Don't forget reverse ethnocentrism.)

19. Explain reentry shock.

20. What viewpoint should one take when studying other cultures?

21. Is cultural relativism equivalent to moral relativism? Explain.

22. What are subcultures? When are subcultures institutionally complete?

Concept Application

Consider the concepts listed below. Match one or more of the concepts with each scenario. Explain your choices.

- ✓ Diffusion
- ✓ Feeling rules
- ✓ Norms
- ✓ Reentry shock
- ✓ Reverse ethnocentrism
- ✓ Subcultures

Scenario 1 "Overseas, the home country environment becomes irrationally glorified. All difficulties and problems are forgotten, and only the good things back home are remembered. Upon returning to the United States, people may be surprised to find that they not only miss their host country and its people, culture, and customs but also the people with whom they shared the experience. They realize how well they actually got along under a different set of living conditions and how much happened and changed back home in their absence. As one woman said to me, 'Three years is a long time to be immersed in another way of life, and I felt numb and kind of left out or not in on things happening in the United States. It was a very unhappy time for me because I had expected to be ecstatic to get home'" (Koehler 1986:90).

Scenario 2 "Yesterday, my four-year-old stopped crying. He fell off his bike, held his breath, and gritted his teeth. 'I'm not gonna cry, Mom,' he said. 'I'm really not.' Where did this pint-size stoicism come from? Batman videos? Preschool name-callers? Maybe the neighbors who tell their kid, 'Crying will get you nowhere.' You hear it everywhere. You'd better not pout; you'd better not cry. Big boys don't cry. Grin and bear it, hide it, stifle it, but whatever you do, don't cry. *Please*, don't cry. I'll give you a cookie if you stop" (Hogan 1994:E1).

Scenario 3
Teens who embrace Goth "celebrate the darker side of humanity, with most young people wearing black clothing, pale makeup with dark accents, and jet-black hair styled in an unusual manner, though it's much more than appearance…American Goths wear piercings and tattoos across their faces and bodies. Goths sometimes pierce their lips, foreheads, and eyebrows, as well as their ear lobes" (Brooks 2007).

Scenario 4
"Some of the world's top donut chains have come rolling into China, Taiwan, South Korea and Japan, and elsewhere in the region as Asians embrace the Western fast food fad. Chains like Krispy Kreme, Dunkin' Donuts, and Mister Donut are setting up shop in a region not known for its sweet tooth, reflecting a growing openness to foreign foods and rising living standards according to the chains and consumers who sometimes wait in line for hours for the treats" (Young 2007).

Scenario 5 "Japanese frequently bow to one another—for instance, when greeting someone—as a gesture of respect and sincerity. The type of bow depends on the formality of the situation, the type of personal relationship (e.g., close or distant), and the differences in social status of the individuals involved. The bow might be no more than a simple nod of the head or, on more formal occasions, a deeper bow from the waist. The most formal bow involves kneeling, placing one's hands out in front on the floor, and lowering the head slowly so that it almost touches the floor. Bowing is not always required, however. Family members and close friends do not usually bow to each other, but a child might bow to his or her mother when apologizing for mischievous behavior" (Japan Information Center 1988: 61).

Practice Test: Multiple-Choice Questions

1. Currently there are about _____ U.S. military personnel stationed in South Korea.
    a. 10,000
    b. 30,000
    c. 83,000
    d. 100,000

2. Which one of the following descriptions applies to South Korea?
    a. communist-style government
    b. isolated
    c. centrally-planned economy
    d. top 20 economy

3. Sociologists face a number of challenges in studying culture? Those challenges include all but which one of the following?
    a. describing culture
    b. determining who belongs to a culture
    c. identifying the distinguishing characteristics that set one culture apart from another
    d. identifying overlap among cultures

4. American sociologists studying Korean bathhouses would be struck by the
    a. private nature of the bath.
    b. tense atmosphere.
    c. lack of self-consciousness regarding the body.
    d. casual relationships between adult men and women.

5. A sociologist seeking to *explain* why Koreans work harder to save energy than Americans would explore the role of
    a. genes.
    b. norms.
    c. values.
    d. geographic and historical factors.

6. The value underlying Korean use of "our" versus "my" is
    a. survival of the fittest.
    b. the self-made person.
    c. the importance of the group.
    d. individual achievement.

7. North Korean president Kim Il-Sung was raised a Christian and even played the church organ. After taking power, Kim completely wiped out Christianity from his country. This example supports the view that
    a. people are cultural replicas of one another.
    b. people have the power to reject, manipulate, and create culture.
    c. there are cultural formulas for passing on cultural experiences.
    d. people are passive agents who absorb one version of culture.

8. One indicator of culture's influence on satisfying hunger is that
    a. only a portion of the potential food available is defined as edible.
    b. people everywhere eat three meals a day.
    c. fast food appeals to people everywhere.
    d. if people are hungry enough, they will eat just about anything.

9. _____ is a staple of the American diet.
    a. Hamburgers
    b. Hot dogs
    c. Rice
    d. Corn

10. _____ are internal bodily sensations that we experience in relationships with other people.
    a. Social emotions
    b. Feeling rules
    c. Emotional states
    d. Expressive norms

11. The U.S. Army publishes a list of "Must Know Items" about South Korea for American soldiers who are stationed there. One item says, "Don't be surprised if you see two Korean women or men walking arm in arm. They are just good friends, and there is nothing sexual implied." The Army is alerting soldiers to
    a. material culture.
    b. feeling rules.
    c. reverse ethnocentrism.
    d. idioms.

12. The 86,000 South Korean Jehovah's Witnesses trace the roots of their religion to _____, the home of the *Watchtower* publication.
    a. Pittsburgh, Pennsylvania
    b. New York, New York
    c. Chicago, Illinois
    d. Mexico City

13. The North Korean government prohibits its 23 million people from receiving mail or telephone calls from outside the country. By doing this, the North Korean government is
    a. creating a cultureless society.
    b. severely limiting opportunities for cultural diffusion.
    c. supporting cultural relativism.
    d. introducing culture shock.

14. _____ is the strain that people from one culture experience when they must reorient themselves to the ways of a new culture.
    a. Culture shock
    b. Ethnocentrism
    c. Diffusion
    d. Reverse ethnocentrism

15. The tendency to hold your own culture as a standard against which other cultures are judged is
    a. cultural relativity.
    b. cultural awareness.
    c. ethnocentrism.
    d. multicultural relativism.

16. Reentry shock is _____ in reverse; it is experienced upon returning home after living in another culture.
    a. material culture
    b. culture shock
    c. ethnocentrism
    d. cultural relativity

17. Under Japanese rule, Korean students were taught by Japanese teachers, Korean names were changed to Japanese names, and practically everything Korean was abandoned. The Japanese were guilty of
    a. cultural relativity.
    b. institutional completeness.
    c. reverse ethnocentrism.
    d. cultural genocide.

18. An individual who adopts cultural relativism aims to _____ a cultural practice.
    a. understand
    b. condone
    c. discredit
    d. accept uncritically

19. The 29,600 American men and women stationed in South Korea spend most of their time at one of 46 military bases. This suggests that the military in South Korea is
    a. institutionally complete.
    b. promoting cultural diffusion.
    c. encouraging cultural immersion.
    d. advocating reverse ethnocentrism.

20. Which one of the following groups represents an example of a counterculture?
    a. a sorority
    b. a fraternity
    c. a retirement community
    d. Buddhist monks

True/False Questions

1. T F U.S. servicemen and servicewomen have fought, died, and otherwise served in Korea to unite the North and South.

2. T F South Korea possesses one of the top 20 economies in the world.

3. T F Much of the Korean identity is intricately linked with the idea of being "not Japanese."

4. T F The opportunity for cultural diffusion occurs whenever people from different cultures make contact.

5. T F Some South Koreans have "borrowed" the religion of the Jehovah's Witnesses, which originated in the United States.

6. T F The North Korean government encourages cultural diffusion.

7. T F Travelers are more likely to prepare for the experience of culture shock than to prepare for reentry shock.

8. T F After Japan annexed Korea in 1910, Japanese became the official language.

9. T F For the most part, the 23 million people of North Korea are not permitted to travel beyond their country's borders.

10. T F Reentry shock is the glorification of a home country.

Internet Resources Related to South and North Korea

- **Korea Herald**
  http://www.koreaherald.co.kr
  Korea Herald is major newspaper of Korea. Read this paper to keep up with latest news from Korea.

- **The Korean Society**
  www.kimsoft.com/korea.htm
  The Korean Society is "dedicated solely to the promotion of greater awareness, understanding, and cooperation between the people of the United States and Korea."

Applied Research

Find a Handbook for International Students online. Typically, these handbooks include a section that prepares international students for life in the United States by giving general descriptions of American culture (dating, friendships, work ethic, beliefs, values, and so on). Use the information in the handbook to create a pocket-sized card highlighting "do's and don'ts" for living in the United States. One example of a handbook can be found at the University of Minnesota Web site: www.isss.umn.edu/new/HandBook/6.pdf.

Size: about the size of Indiana
Capital: Seoul (10.3 million)
Population (2006): 48,846,823
Ethnic groups: Korean; small Chinese minority
Religions: Christianity, Buddhism, Shamanism, Confucianism, Chondogyo
Infant mortality rate: 6.16/1,000
Life expectancy: 77.0 yrs (males 73.6 yrs., females 80.8 yrs)
Work force (2005): 23.53 million
Workforce by Sector: Services, 67.2%; mining and manufacturing, 26.4%; agriculture, 6.4%
Government Type: Republic with powers shared between the president, the legislature, and the courts

Korea's population is one of the most ethnically and linguistically homogenous in the world. Except for a small Chinese community (about 20,000), virtually all Koreans share a common cultural and linguistic heritage. With 48.85 million people, South Korea has one of the world's highest population densities

Korea has experienced one of the largest rates of emigration, with ethnic Koreans residing primarily in China (1.9 million), the United States (1.52 million), Japan (681,000), and the countries of the former Soviet Union (450,000).

Half of the population actively practices religion. Among this group, Christianity (49%) and Buddhism (47%) comprise Korea's two dominant religions. Though only 3 percent identified themselves as Confucianists, Korean society remains highly imbued with Confucian values and beliefs. The remaining 1 percent of the population practices Shamanism (traditional spirit worship) and Chondogyo ("Heavenly Way"), a traditional religion.

The Republic of Korea's economic growth over the past 30 years has been spectacular. Per capita GNP, only $100 in 1963, exceeded $16,000 in 2005. South Korea is now the United States' seventh largest trading partner and is the eleventh largest economy in the world.

North and South Korea have moved forward on a number of economic cooperation projects. The most prominent projects include (1) the construction of the Kaesong Industrial Complex, (2) tourism, and (3) infrastructure development. (1) The Kaesong Industrial Complex (KIC), since its June 2003 groundbreaking, has grown to include a variety of South Korean companies operating in this North-South cooperation project. The R.O.K. envisages a substantial enlargement of participation in the project in the following years, although new investment was suspended following the North's testing of a nuclear device in October 2006. (2) Tourism: R.O.K.-organized tours to Mt. Kumgang in North Korea began in 1998. Since then, more than a million visitors have traveled to Mt. Kumgang. (3) Infrastructure Development: Although east and west coast railroad and roads links have been reconnected across the DMZ, neither rail link has been tested. The roads crossing the DMZ are used on a daily basis between South Korea and Mt. Kumgang, as well as to the Kaesong Industrial Complex.

Two-way trade between North and South Korea, legalized in 1988, hit almost $1.35 billion in 2006, up 27.8 percent from 2005. This total included a substantial quantity of non-trade goods provided to the North as aid (fertilizer, etc.) or as part of inter-Korean cooperative projects. According to R.O.K. figures, about 60 percent of the total trade consisted of commercial transactions, much of that based on processing-on-commission arrangements. The R.O.K. is North Korea's second-largest trading partner.

Source: Excerpted from U.S. Department of States, *Background Notes*
www.state.gov/r/pa/ei/bgn/2800.htm#relations

Size: about the size of Mississippi
Capital: Pyongyang
Terrain: About 80 percent of land area is moderately high mountains separated by deep, narrow valleys and small, cultivated plains. The remainder is lowland plains covering small, scattered areas.
Population (2006): 23.1 million
Ethnic groups: Korean; small ethnic Chinese and Japanese populations
Religions: Buddhism, Confucianism, Shamanism, Chongdogyo, Christian; autonomous religious activities have been virtually nonexistent since 1945
Infant mortality rate: 23.29/1,000
Life expectancy: Males 68 yrs., females 74 yrs. (2006 est.)
Government Type:

North Korea now has the fourth largest army in the world. It has an estimated 1.21 million armed personnel, compared to about 680,000 in the South. Military spending is estimated at as much as a quarter of GNP, with about 20 percent of men ages 17 to 54 in the regular armed forces. North Korean forces have a substantial numerical advantage over the South (between 2 and 3 to 1) in several key categories of offensive weapons—tanks, long-range artillery, and armored personnel carriers. The North has perhaps the world's second largest special operations force, designed for insertion behind the lines in wartime. While the North has a relatively impressive fleet of submarines, its surface fleet has a very limited capability. Its air force has twice the number of aircraft as the South, but except for a few advanced fighters, the North's air force is obsolete. The North deploys the bulk of its forces well forward, along the demilitarized zone (DMZ). Several North Korean military tunnels under the DMZ were discovered in the 1970s.

44

Over the last several years, North Korea has moved more of its rear-echelon troops to hardened bunkers closer to the DMZ. Given the proximity of Seoul to the DMZ (some 25 miles), South Korean and U.S. forces are likely to have little warning of any attack. The U.S. and South Korea continue to believe that the U.S. troop presence in South Korea remains an effective deterrent. North Korea's nuclear weapons program has also been a source of international tension.

North Korea's economy declined sharply in the 1990s with the end of communism in Eastern Europe, the disintegration of the Soviet Union, and the dissolution of bloc-trading with the countries of the former socialist bloc. Gross national income per capita is estimated to have fallen by about one-third between 1990 and 2002. The economy has since stabilized and has shown some modest growth in recent years, which may be reflective of increased inter-Korean economic cooperation. Still, output and living standards remain far below 1990 levels. Other centrally-planned economies in similar straits opted for domestic economic reform and liberalization of trade and investment. North Korea formalized some modest wage and price reforms in 2002 and has increasingly tolerated markets and a small private sector as the state-run distribution system has deteriorated. But the regime seems determined to maintain control. In October 2005, emboldened by an improved harvest and increased food donations from South Korea, the North Korean government banned private grain sales and announced a return to centralized food rationing. Reports indicate that this effort to reassert state control and to control inflation has been largely ineffective. Another factor contributing to the economy's poor performance is the disproportionately large share of GDP (thought to be about one-fourth) that North Korea devotes to its military.

North Korean industry is operating at only a small fraction of capacity due to lack of fuel, spare parts, and other inputs. Agriculture is now 30 percent of GDP, even though agricultural output has not recovered to early 1990 levels. The infrastructure is generally poor and outdated, and the energy sector has collapsed. About 80 percent of North Korea's terrain consists of moderately high mountain ranges and partially forested mountains and hills separated by deep, narrow valleys and small, cultivated plains. The most rugged areas are the north and east coasts. Good harbors are found on the eastern coast. Pyongyang, the capital, which is situated near the country's west coast, is located on the Taedong River.

North Korea experienced a severe famine following record floods in the summer of 1995 and continues to suffer from chronic food shortages and malnutrition. The United Nations World Food Program (WFP) provided substantial emergency food assistance beginning in 1995 (2 million tons of which came from the U.S.), but the North Korean government suspended the WFP emergency program at the end of 2005. It has since permitted the WFP to resume operations on a greatly reduced scale through a Protracted Relief and Recovery Operation. External food aid now comes primarily from China and South Korea in the form of grants and long-term concession loans. South Korea also donates fertilizer and other materials, while China provides energy. South Korea suspended food and fertilizer shipments to the North in response to North Korea's missile launches in July 2006. Later that month, when severe floods threatened to produce another humanitarian crisis, South Korea announced a one-time donation of 100,000 tons of food, matching contributions from South Korean non-governmental organizations (NGOs). South Korea resumed fertilizer shipments to North Korea in late March 2007.

Legalized in 1988, two-way trade between North and South Korea had risen to more than $1 billion by 2005, much of it related to out-processing or assembly work undertaken by South Korean firms in the Kaesong Industrial Complex (KIC). A significant portion of the total also includes donated goods provided to the North as humanitarian assistance or as part of inter-Korean cooperation projects. Although business-based and processing-on-commission transactions have continued to grow, the bulk of South Korean exports to North Korea in 2006 were still non-commercial.

Since the June 2000 North-South summit, North and South Korea have reconnected their east and west coast railroads and roads where they cross the DMZ, and they are working to improve these transportation routes. To date, the railroads have not been tested. Much of the work done in North Korea has been funded by the South. The west coast rail and road are complete as far north as the KIC (six miles north of the DMZ), but little work is being done north of Kaesong. On the east coast, the road is complete, but the rail line is far from operational. Since 2003, tour groups have been using the east coast road to travel from South Korea to Mt. Geumgang in North Korea, where cruise ship based tours have been permitted since 1998.

As of February 2007, 21 South Korean firms were manufacturing goods in the KIC, employing nearly 9,000 North Korean workers. Most of the goods are sold in South Korea; a small quantity is being exported to foreign markets. Groundbreaking for the KIC was in June 2003, and the first products were shipped from the complex in December 2004. Plans envision 250 firms employing 350,000 workers by 2012. South Korea's Ministry of Unification announced in September 2006 that it was postponing further expansion plans in light of heightened tensions following the North's ballistic missile launches. Despite the September decision, preexisting agreements to build or operate new factories are being executed.

The U.S. imposed a near total economic embargo on North Korea in June 1950 when North Korea attacked the South. Sanctions were eased in stages beginning in 1989 and following the Agreed Framework on North Korea's nuclear programs in 1994. In June 2000, a U.S. Executive Order legalized most transactions between U.S. and North Korean persons. It allowed most products, other than those specifically controlled for military, non-proliferation, or anti-terrorism purposes, to be exported to North Korea without an export license. Restrictions on U.S. investments in North Korea and travel of U.S. citizens to North Korea were also eased, and U.S. ships and aircraft were allowed to call at North Korean ports. To date, U.S. economic interaction with North Korea remains minimal, licenses are still required for imports from North Korea, and North Korean assets frozen since 1950 remain frozen. In January 2007, pursuant to UN Security Council Resolution 1718, the U.S. Department of Commerce issued new regulations prohibiting the export of luxury goods to North Korea. Many statutory sanctions on North Korea, including those affecting trade in military, dual-use, and missile-related items and those based on multilateral arrangements, remain in place. Most forms of U.S. economic assistance, other than purely humanitarian assistance, are prohibited. North Korea does not enjoy "Normal Trade Relations" with the U.S., so any goods manufactured in North Korea are subject to a higher tariff upon entry to the United States.

**Source: Excerpted from U.S. Department of States, *Background Notes*
http://www.state.gov/r/pa/ei/bgn/2792.htm**

Answers

Concept Application

1. Reverse ethnocentrism;
   reentry shock
2. Feeling rules
3. Subculture
4. Diffusion
5. Norms

Multiple-Choice		**True/False**
1. b page 71	11. b page 84	1. F page 85
2. d page 72	12. a page 85	2. T page 75
3. d page 74	13. b page 87	3. F page 83
4. c page 75	14. a page 88	4. T page 76
5. d page 77	15. c page 88	5. T page 87
6. c page 77	16. c page 92	6. F page 84
7. b page 81	17. d page 92	7. T page 90
8. a page 78	18. a page 96	8. T page 95
9. d page 81	19. a page 96	9. T page 96
10. a page 81	20. d page 100	10. F page 92

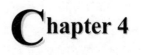hapter 4

Socialization

Study Questions

1. Why are Israel and the Palestinian Territories paired with the concept socialization?

2. What is socialization?

3. Distinguish between nature and nurture.

4. How do extreme cases of isolation underscore the importance of socialization? Choose one of the following cases to illustrate: (a) Anna and Isabelle; (b) children orphaned as a result of the Holocaust; (c) Spitz's study of orphanages for children of prison mothers; and (d) the elderly in nursing homes.

5. On the basis of Anna and Isabelle's case histories, what conclusions did Kingsley Davis reach about the effects of prolonged isolation?

6. What are the basic dynamics underlying the century-long struggle between Palestinian Arabs and Jews? What kinds of issues must be resolved if the peace process is to move forward?

7. What is the social importance of memory?

8. Define collective memory. How is collective memory passed on?

9. What are significant symbols?

10. Distinguish between the "I" and the "me." How does the "me" develop?

11. What is role-taking? What are the stages by which children come to learn to take the role of others?

12. According to Charles Horton Cooley's "Looking Glass-Self" theory, how does a sense of self develop?

13. What central concept underlies Piaget's theory of cognitive development? What are the four stages of cognitive development?

14. What are agents of socialization?

15. What are primary groups? How are they important agents of socialization?

16. What characteristics make a military unit a primary group?

17. What are ingroups and outgroups? What is the sociological significance of ingroups and outgroups?

18. How does Emile Durkheim define suicide? What are the four types?

19. How might Durkheim classify the ties that bind Palestinian suicide bombers/martyrs to the group?

20. What is the mass media? How does it affect the sense of self and individuals' relationships to others?

21. Why is Sesame Street called Sesame Stories in Israel and the Palestinian Terrorities?

22. Briefly summarize the eight stages of life cycle.

23. What is resocialization? What are the types of resocialization?

24. What are total institutions, and what mechanisms do total institutions use to resocialize inmates?

25. Under what conditions are people least likely to resist resocialization?

Concept Applications

Consider the concepts listed below. Match one or more of the concepts with each scenario. Explain your choices.

    a. Collective memory
    b. Nature
    c. Nurture
    d. Resocialization
    e. Total institutions
    f. Anomic social relationships
    g. Altruistic social relationships

Scenario 1
    "In 1910, two French surgeons wrote about their successful operation on an 8-year-old boy who had been blind since birth because of cataracts. When the boy's eyes were healed, they removed the bandages, eager to discover how well the child could see. Waving a hand in front of the boy's physically perfect eyes, they asked him what he saw. He replied weakly, 'I don't know.' 'Don't you see it moving?' they asked. 'I don't know,' was his only reply. The boy's eyes were clearly not following the slowly moving hand. What he saw was only a varying brightness in front of him. He was then allowed to touch the hand. As it began to move, he cried out in a voice of triumph: 'It's moving!' He could feel it move, and even, as he said, 'hear it move,' but he still needed laboriously to learn to see it move" (Zajonc 1993:22).

Scenario 2
    "Let me say to you, the Palestinians, we are destined to live together on the same soil in the same land. We the soldiers who have returned from battles stained with blood; we who have seen our relatives and friends killed before our eyes; we who have attended their funerals and cannot look into the eyes of their parents; we who have come from a land where parents bury their children; we who have fought against you, the Palestinians" (Rabin 1993:A7).

Scenario 3

"Genetic endowments may set limits for the height or intelligence that individuals can attain, but their actual height or intelligence also depends upon how they are raised. The increasing height of the American population over the past several generations reflects the change in nutritional conditions and probably the diminution in childhood illnesses more than a genetic selection" (Lidz 1976:40).

Scenario 4

"Hospitals with hundreds, even thousands of inpatients, maintain schedules aimed at ensuring that every patient receives essential care, and the staff must fit the needs and daily activities of dying patients into the hospital's schedule. They tend to require all patients, whether terminal or not, to give up virtually all personal control over the little things that make up their day-to-day lives. The kinds of personal items that can make a big difference, such as your own pillow from home, are often not allowed. Visits by children may be curtailed, and having a pet stay with a dying person is prohibited. Activities, such as walking, eating, bathing, and any physical exercise will proceed according to an established routine" (Anderson 1991:144).

Scenario 5

Some events are experienced by great numbers of people, diverse in interest, age, race, ethnicity, lifestyle and life chances, gender, language, and place, who temporarily become bound together by a historical moment. The January 28, 1986, Space Shuttle Challenger disaster was such a moment. Collectively, the country grieved, and not for the first time. Many still vividly remember—and will quickly confess, when the subject comes up—exactly where they were, what they were doing, and how they felt when they heard about the tragedy. The initial shock was perpetuated by the television replays of the Challenger's final seconds, by the anguished faces of the astronauts' families and other onlookers huddled in disbelief on bleachers at the launch pad, by the news analyses, and then by the official investigation of the Presidential Commission (Vaughn 1996:xi).

Scenario 6

"Around 400 volunteers signed up in Tehran to sacrifice their lives in "occupied Islamic countries" on Wednesday night, inspired by a fatwa from a top hardline cleric giving religious backing to suicide missions. Wednesday's registration session was the latest by a group called the Committee for the Commemoration of Martyrs of the Global Islamic Campaign, which says it has enrolled 35,000 volunteers nationwide for possible attacks since last year… 'As a Muslim, it is my duty to sacrifice my life for oppressed Palestinian children,' said Maryam Partovi, 31, a mother of two. A banner hanging over the main entrance quoted Khamenei as saying, 'Sacrificing oneself for religion and national interest is the height of honour and bravery.'"

Practice Test: Multiple-Choice Questions

1. The process by which people take as their own and accept as binding the norms, values, beliefs, and language needed to participate in the larger community is termed
    a. adaptation.
    b. internalization.
    c. assimilation.
    d. acculturation.

2. In studying the Israeli-Palestinian conflict, sociologists ask all <u>but</u> which one of the following questions?
    a. How do members learn about and come to terms with the environment they have inherited?
    b. How is conflict passed down from one generation to another?
    c. What roles do nature and nurture play in creating a "Palestinian" and an "Israeli" identity?
    d. Why can't we all just get along?

3. _____ is the term for human genetic makeup or biological inheritance.
    a. Nature
    b. Nurture
    c. Internalization
    d. Socialization

4. The _____ allows us to organize, remember, communicate, understand, and create.
    a. cortex
    b. cerebral cortex
    c. left side of the brain
    d. right side of the brain

5. The cases of Anna and Isabelle were used to illustrate
    a. the importance of social contact for normal development.
    b. the fact that humans are born with a great learning capacity.
    c. that people are born with preconceived notions about standards of appearance and behavior.
    d. that two-year-olds are bothered when rules are violated.

6. _____ control(s) the West Bank, and _____ control(s) the Gaza Strip.
    a. Fatah; Hamas
    b. Kibbutz; Engrams
    c. Hamas; Fatah
    d. Palestinians; Israelis

7. Which one of the following scenarios indicates that a bond of mutual expectation has developed between a baby and his or her caretaker?
    a. A baby learns to comfort itself when it feels distress.
    b. A baby cannot tell how its mother will react to its cries.
    c. A baby sleeps through the night.
    d. A baby comes to expect that if it cries, a caretaker will offer comfort.

8. A cultural center located in the West Bank has established a memorial honoring the lives of Palestinians killed in the second *intifada*. The memorial serves as a vehicle for instilling
    a. active adaptation.
    b. reflective thinking.
    c. engrams.
    d. collective memory.

9. According to George Herbert Mead, the "me" is the part of the self that
    a. is spontaneous and creative.
    b. acts in unconventional ways.
    c. develops through imitation, play, and games.
    d. is capable of rejecting expectations.

10. Language, facial expressions, tone of voice, and posture are all examples of
    a. socialization.
    b. significant symbols.
    c. reflexive thinking.
    d. role taking.

11. Tyler sits at the computer, mimicking the behavior of her mother. Tyler's family encourages this by telling her how cute she is. Tyler is in the
    a. preparatory stage.
    b. play stage.
    c. game stage.
    d. looking-glass self stage.

12. During the game stage, children learn
    a. to make up rules as they go.
    b. to mimic and imitate people in their environment.
    c. to pretend to be people significant in their lives.
    d. to see how their position fits relative to all other positions.

13. An Israeli soldier observes that some Palestinian girls are frightened to death when he is standing in their path. The soldier is engaging in
    a. socialization.
    b. internalization.
    c. role-taking.
    d. collective memory.

14. Which one of the following thinkers is associated with the concept of "active adaptation"?
    a. Erving Goffman
    b. Charles Horton Cooley
    c. George H. Mead
    d. Jean Piaget

15. Significant others, primary groups, ingroups and outgroups, and institutions, such as mass media, are known as
    a. scapegoats.
    b. the looking-glass self.
    c. significant symbols.
    d. agents of socialization.

16. Those groups with which people identify and to which they feel closely attached, particularly when that attachment is founded on hatred of another group, are
    a. essential groups.
    b. respected groups.
    c. outgroups.
    d. ingroups.

17. Almost every Israeli can claim membership in which one of the following primary groups?
    a. a sports team
    b. a college sorority
    c. the Jewish faith
    d. a military unit

18. _____ describes a state in which there is no hope of change.
    a. Anomic
    b. Fatalistic
    c. Altruistic
    d. Egoistic

19. Which area of the world is most likely to have the highest Internet penetration rates?
    a. North America
    b. Middle East
    c. Africa
    d. Asia

20. _____ is United States city with the largest Jewish population in the world.
    a. Miami
    b. New York
    c. San Francisco
    d. Boston

True/False Questions

1. T F   The Israeli-Palestinian conflict has lasted for 20 years.

2. T F   In the first weeks of life, babies are able to babble the sounds needed to speak any language.

3. T F   The political party Fatah is backed by Syria and Iran.

4. T F   Language is a particularly important significant symbol.

5. T F   During the game stage, children learn to take the role of the significant other.

6. T F   If their primary groups remain intact, children can emerge from widespread turmoil, violence, and destruction in relatively good psychological condition.

7. T F   From a sociological point of view, suicide is the severing of relationships.

8. T F   More than 80 percent of suicide bombers/martyrs are unmarried.

9. T F   If all goes well, in Stage 8 (Old Age), a person comes to accept the life he or she has lived.

10. T F   Mental hospitals, concentration camps, and boarding schools are primary groups.

Internet Resources

- **News Summaries from Today's Israeli Papers**
  http://www.israelemb.org/useful_links.html
  The Embassy of Israel – Washington, D.C. translates articles from various Israeli dailies including Ma'ariv and HaAretz.

- *The Jerusalem Report*
  http://www.jrep.com/
  The Jerusalem Report is a bi-weekly publication that "brings the most thorough and in-depth analysis of the controversial issues facing Israel today."

- **Palestinian Central Bureau of Statistics**
  http://www.pcbs.org/
  The Palestinian Central Bureau of Statistics has posted a variety of information on Palestine including population and economic statistics. There is also a list of links to websites such as Palestine Facts, Palestinian Authority and Palestine Colleges and Universities.

Applied Research

We learned in Chapter 4 that resocialization is the process of being socialized over again. In particular it is a process of discarding values and behaviors unsuited to new circumstances and replacing them with new, more appropriate values and norms. Identify a resocialization situation—boot camp, first days on a new job, graduation, self-help group, and so on. Use the internet, library research, or personal interviews with someone who has been resocialized to identify critical stages and key events in the resocialization process.

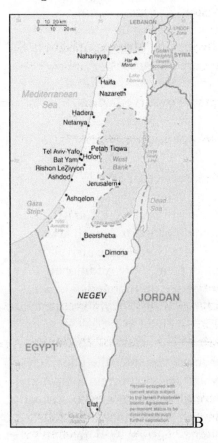

Israel

Size: About the size of New Jersey
Capital: Jerusalem
Population (2005): 6.35 million
Ethnic groups: Jews, 76.2%; Arabs, 19.5%; other, 4.3%
Religions: Judaism, Islam, Christianity, Druze
Infant mortality rate: 21.9/1,000 births
Life expectancy at birth: female 81.55 years; male 77.21 years
Work force: 2.68 million
Workforce by Sector: Agriculture, 2.1%; manufacturing, 16.2%; construction, 5.4%; business activities, 13.4%; education, 12.7%; health, welfare, and social services, 10.7%;

Gaza

Size: About twice the size of Washington, DC
Capital: Jerusalem
Population (2007): 1.48 million
Ethnic groups: Palestinian Arab and other 99.4%, Jewish 0.6%
Religions: Muslim (predominantly Sunni) 98.7%, Christian 0.7%, Jewish 0.6%
Infant mortality rate: 7.03/1,000 births
Life expectancy at birth: female 74.0 years; male 71.0 years
Work force: 259,000
Workforce by Sector: Agriculture, 12.1%; industry, 18.0%; services, 70.0

West Bank

Size: slightly smaller than Delaware
Capital: Jerusalem
Population (2007): 2,535,927; in addition, there are about 187,000 Israeli settlers in the West Bank and fewer than 177,000 in East Jerusalem
Ethnic groups: Palestinian Arab and other 83%, Jewish 17%
Religions: Muslim 75% (predominantly Sunni), Jewish 17%, Christian and other 8%
Infant mortality rate: 18.7/1,000 births
Life expectancy at birth: female 75.4 years; male 71.7 years
Work force: 568,000
Workforce by Sector: Agriculture, 8.0%; industry, 29.0%; services, 55.0%

Of the approximately 6.35 million Israelis in 2005, about 4.86 million were counted as Jewish, though some of those are not considered Jewish under Orthodox Jewish law. Since 1989, nearly a million immigrants from the former Soviet Union have arrived in Israel, making this the largest wave of immigration since independence. Additionally, almost 50,000 members of the Ethiopian Jewish community have immigrated to Israel, 14,000 of them during the dramatic May 1991 Operation Solomon airlift. Approximately 35 percent (35.3%) of Israelis were born outside of Israel.

The three broad Jewish groupings are the Ashkenazim, or Jews who trace their ancestry to western, central, and eastern Europe; the Sephardim, who trace their origin to Spain, Portugal, southern Europe, and North Africa; and Eastern or Oriental Jews, who descend from ancient communities in Islamic lands. Of the non-Jewish population, about 68 percent are Muslims, about 9 percent are Christian, and about 7 percent are Druze.

The creation of the State of Israel in 1948 was preceded by more than 50 years of efforts to establish a sovereign nation as a homeland for Jews. These efforts were initiated by Theodore Herzl, founder of the Zionist movement and were given added impetus by the Balfour Declaration of 1917, which asserted the British Government's support for the creation of a Jewish homeland in Palestine.

In the years following World War I, Palestine became a British Mandate, and Jewish immigration steadily increased, as did violence between Palestine's Jewish and Arab communities. Mounting British efforts to restrict this immigration were countered by international support for Jewish national aspirations following the near-extermination of European Jewry by the Nazis during World War II. This support led to the 1947 UN partition plan, which would have divided Palestine into separate Jewish and Arab states, with Jerusalem under UN administration.

On May 14, 1948, soon after the British quit Palestine, the State of Israel was proclaimed and was immediately invaded by armies from neighboring Arab states, which rejected the UN partition plan. This conflict, Israel's War of Independence, was concluded by armistice agreements between Israel, Egypt, Jordan, Lebanon, and Syria in 1949 and resulted in a 50 percent increase in Israeli territory.

In 1956, French, British, and Israeli forces engaged Egypt in response to its nationalization of the Suez Canal and blockade of the Straits of Tiran. Israeli forces withdrew in March 1957, after the United Nations established the UN Emergency Force (UNEF) in the Gaza Strip and Sinai. This war resulted in no territorial shifts and was followed by several years of terrorist incidents and retaliatory acts across Israel's borders.

In June 1967, Israeli forces struck targets in Egypt, Jordan, and Syria in response to Egyptian President Nasser's ordered withdrawal of UN peacekeepers from the Sinai Peninsula and the buildup of Arab armies along Israel's borders. After six days, all parties agreed to a cease-fire, under which Israel retained control of the Sinai Peninsula, the Golan Heights, the Gaza Strip, the formerly Jordanian-controlled West Bank of the Jordan River, and East Jerusalem. On November 22, 1967, the Security Council adopted Resolution 242, the "land for peace" formula, which called for the establishment of a just and lasting peace based on Israeli withdrawal from territories occupied in 1967 in return for the end of all states of belligerency, respect for the sovereignty of all states in the area, and the right to live in peace within secure, recognized boundaries.

The following years were marked by continuing violence across the Suez Canal, punctuated by the 1969 and 1970 war of attrition. On October 6, 1973, Yom Kippur (the Jewish Day of Atonement), the armies of Syria and Egypt launched an attack against Israel. Although the Egyptians and Syrians initially made significant advances, Israel was able to push the invading armies back beyond the 1967 cease-fire lines by the time the United States and the Soviet Union helped bring an end to the fighting. In the UN Security Council, the United States supported Resolution 338, which reaffirmed Resolution 242 as the framework for peace and called for peace negotiations between the parties.

In the years that followed, sporadic clashes continued along the cease-fire lines but guided by the U.S., Egypt and Israel continued negotiations. In November 1977, Egyptian President Anwar Sadat made a historic visit to Jerusalem, which opened the door for the 1978 Israeli-Egyptian peace summit convened at Camp David and hosted by President Carter. These negotiations led to a 1979 peace treaty between Israel and Egypt, pursuant to which Israel withdrew from the Sinai in 1982, signed by President Sadat of Egypt and Prime Minister Menahem Begin of Israel.

In the years following the 1948 war, Israel's border with Lebanon was quiet relative to its borders with other neighbors. After the expulsion of Palestinian fighters from Jordan and their influx into southern Lebanon in 1970, hostilities along Israel's northern border increased, and Israeli forces crossed into Lebanon. After the passage of Security Council Resolution 425, which called for Israeli withdrawal and the creation of the UN Interim Force in Lebanon peacekeeping force (UNIFIL), Israel withdrew its troops.

In June 1982, following a series of cross-border terrorist attacks and the attempted assassination of the Israeli Ambassador to the U.K., Israel invaded Lebanon to fight the forces of Yasser Arafat's Palestine Liberation Organization (PLO). The PLO withdrew its forces from Lebanon in August 1982. Israel, having failed to finalize an agreement with Lebanon, withdrew most of its troops in June 1985 save for a residual force that remained in southern Lebanon to act as a buffer against attacks on northern Israel. These remaining forces were completely withdrawn in May 2000 behind a UN-brokered delineation of the Israel-Lebanon border (the Blue Line). Hizballah forces in Southern Lebanon continued to attack Israeli positions south of the Blue Line in the Sheba Farms/Har Dov area of the Golan Heights.

The victory of the U.S.-led coalition in the Persian Gulf War of 1991 opened new possibilities for regional peace. In October 1991, the United States and the Soviet Union convened the Madrid Conference, in which Israeli, Lebanese, Jordanian, Syrian, and Palestinian leaders laid the foundations for ongoing negotiations designed to bring peace and economic development to the region. Within this framework, Israel and the PLO signed a Declaration of Principles on September 13, 1993, which established an ambitious set of objectives relating to a transfer of authority from Israel to an interim Palestinian authority. Israel and the PLO subsequently signed the Gaza-Jericho Agreement on May 4, 1994, and the Agreement on Preparatory Transfer of Powers and Responsibilities on August 29, 1994, which began the process of transferring authority from Israel to the Palestinians.

On October 26, 1994, Israel and Jordan signed a historic peace treaty, which was witnessed by President Clinton. This was followed by Israeli Prime Minister Rabin and PLO Chairman Arafat's signing of the historic Israeli-Palestinian Interim Agreement on September 28, 1995. This accord, which incorporated and superseded previous agreements, broadened Palestinian self-government and provided for cooperation between Israelis and Palestinians in several areas.

Israeli Prime Minister Yitzhak Rabin was assassinated on November 4, 1995, by a right-wing Jewish radical, bringing the increasingly bitter national debate over the peace process to a climax. Subsequent Israeli governments continued to negotiate with the PLO, resulting in additional agreements, including the Wye River and the Sharm el-Sheikh memoranda. However, a summit hosted by President Clinton at Camp David in July 2000 to address permanent status issues, (including the status of Jerusalem, Palestinian refugees, Israeli settlements in the West Bank and Gaza, final security arrangements, borders, and relations and cooperation with neighboring states) failed to produce an agreement.

Following the failed talks, widespread violence broke out in Israel, the West Bank, and Gaza in September 2000. In April 2001, the Sharm el-Sheikh Fact Finding Committee, commissioned by the October 2000 Middle East Peace Summit and chaired by former U.S. Senator George Mitchell, submitted its report, which recommended an immediate end to the violence followed by confidence-building measures and a resumption of security cooperation and peace negotiations. Building on the Mitchell report, in April 2003, the Quartet (the U.S., UN, European Union (EU), and the Russian Federation) announced the "roadmap," a performance-based plan to bring about two states, Israel and a democratic, viable Palestine, living side by side in peace and security.

Despite the promising developments of spring 2003, violence continued and in September 2003 the first Palestinian Prime Minister, Mahmoud Abbas (Abu Mazen), resigned after failing to win true authority to restore law and order, fight terror, and reform Palestinian institutions. In response to the deadlock, in the 2003/2004 winter, Prime Minister Sharon put forward his Gaza disengagement initiative, proposing the withdrawal of Israeli settlements from Gaza as well as parts of the northern West Bank. President Bush endorsed this initiative in an exchange of letters with Prime Minister Sharon on April 14, 2004, viewing Gaza disengagement as an opportunity to move toward implementation of the two-state vision and begin the development of Palestinian institutions. In a meeting in May 2004, the Quartet endorsed the initiative, which was approved by the Knesset in October 2004.

The run-up to disengagement saw a flurry of diplomatic activity, including the February 2005 announcement of Lieutenant General William Ward as U.S. Security Coordinator; the March 2005 Sharon-Abbas summit in Sharm el-Sheikh; the return of Egyptian and Jordanian ambassadors to Israel; and the May 2005 appointment of former World Bank president James D. Wolfensohn as Special Envoy for Gaza Disengagement to work for a revitalization of the Palestinian economy after disengagement. Wolfensohn's direct involvement spurred Israeli-Palestinian agreement on the Gaza crossings at Karni and Erez, on the demolition of settler homes, water, electricity, and communications infrastructure issues, as well as other issues related to the Palestinian economy.

On August 15, 2005, Israel began implementing its disengagement from the Gaza Strip, and the Israeli Defense Forces completed their withdrawal, including the dismantling of 17 settlements, on September 12. After broad recognition for Prime Minister Sharon's accomplishment at that fall's UN General Assembly, international attention quickly turned to efforts to strengthen Palestinian governance and the economy in Gaza. The United States brokered a landmark Agreement on Movement and Access between the parties in November 2005 to facilitate further progress on Palestinian economic issues. However, the terrorist organization Hamas—building on popular support for its "resistance" to Israeli occupation and a commitment to clean up the notorious corruption of the Palestinian Authority—took a majority in the January 2006 Palestinian legislative elections. The Israeli leadership pledged not to work with a Palestinian government in which Hamas had a role, while the Quartet made clear that to receive international support, a new Palestinian government would have to renounce terror and violence, recognize Israel, and accept previous obligations and agreements, including the roadmap. The new government rejected these principles when it was seated on April 1, 2006.

Source: Excepted from U.S. Department of State: Background Notes
http://www.state.gov/r/pa/ei/bgn/35749.htm

Answers

Concept Applications
1. Nature, Nurture, Resocialization
2. Collective memory
3. Nature, Nurture
4. Resocialization, Total institutions
5. Collective memory
6. Altruistic social relationship

	Multiple Choice				**True/False**	
1.	b page 85	11.	a page 94	1.	F page 85	
2.	d page 85	12.	d page 96	2.	T page 86	
3.	a page 86	13.	c page 97	3.	F page 89	
4.	b page 86	14.	d page 98	4.	T page 93	
5.	a page 87	15.	d page 100	5.	F page 96	
6.	a page 89	16.	d page 101	6.	T page 101	
7.	d page 90	17.	d page 101	7.	T page 103	
8.	d page 91	18.	b page 103	8.	T page 104	
9.	c page 94	19.	a page 105	9.	T page 108	
10.	b page 93	20.	b page 109	10.	F page 110	

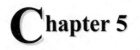hapter 5

Social Interaction

Study Questions

1. Why is the Democratic Republic of the Congo the country of emphasis for the topic of social interaction?

2. What is social interaction? How do sociologists approach the study of social interaction?

3. How did Durkheim define the division of labor? How is the division of labor related to colonization?

4. How did the Democratic Republic of Congo, especially Leopoldville, become part of the global economy?

5. Distinguish between organic and mechanical solidarity.

6. What kinds of disruptions to the division of labor break down the abilities of people to connect with one another in meaningful ways through their labor?

7. How did the European-imposed forced division of labor contribute to the origin of AIDS/HIV?

8. How is the activation and spread of HIV connected to the unprecedented mixing of people from all over the world?

9. Who is responsible for triggering and transmitting HIV? Explain your answer.

10. What is a social status? How is it related to social structure?

11. How are the concepts of status, role, rights, and obligations related?

12. Distinguish between ascribed, achieved, and master statuses?

13. Distinguish between role strain and role conflict?

14. Does the idea of role imply totally predictable behavior? Explain.

15. What are the broad differences between Africa and Congolese regarding the causes and treatment of diseases?

16. What is impression management? What interaction dilemmas are associated with impression management?

17. Apply the concept of impression management to King Leopold and his interest in the Congo.

18. What is the difference between backstage and frontstage? Use these concepts to analyze blood donors' answers to screening questions?

19. People usually attribute cause to either dispositional traits or situational factors. What is the difference between the two factors? Give an example of each.

20. What is a scapegoat? Under what conditions is a person or group likely to be made a scapegoat?

21. What problems are associated with using dispositional traits to explain the cause of AIDS and to diagnose AIDS cases?

22. How does someone get a diagnosis of AIDS? Give an example of how the definition of AIDS shapes understanding of who has AIDS?

23. What must take place before we can truly understand the cause of HIV and AIDS?

Concept Applications

Consider the concepts listed below. Match one or more of the concept with each scenario. Explain your choices.

    a. Dispositional traits
    b. Front stage
    c. Impression management
    d. Role strain
    e. Situational factors
    f. Social interactions

Scenario 1

"Ten minutes after William Andrews succumbed to the poisonous concoction injected into his arm, Dr. Robert Jones performed a task from which he said he would never quite recover. He entered the chamber of death, checked the condemned man's vital signs, and confirmed that he was, in fact, dead.

"The medical director for the Utah State Prison system did not witness the July 1992 execution, but his limited role so troubled him that he decided never again to have anything to do with a state-ordered killing.

'It was much more stressful, much more disconcerting than I though it would be,' Jones says. 'I literally slept for a whole day afterward, and I thought, 'That's an experience in life that you don't want to have to go through again… Physicians usually try to preserve life, not end it.'

"As a prison doctor, Jones sits at the uncomfortable intersection of medicine and criminal justice. His dilemma highlights an ethical debate that is raging in the medical community: should doctors, who take the Hippocratic Oath not to harm their patients, take part in carrying out the death penalty? When state laws and regulations require physicians to be present at executions—as in California, where doctors watch the heart monitor that charts the prisoner's final moments in the gas chamber—should the physician comply?" (Stolberg 1994:E1).

Scenario 2

"Janet's sister, Pam, and brother, Nicholas, along with their own spouses and children, had a hard time understanding what was happening to their mother. It took them longer than it took Janet to catch on because their mother managed to do a superb job of keeping up appearances during the quarter of an hour or so each week when they spoke with her on the phone. And because they didn't want anything major to be the matter either, they weren't able to take Janet's worrying seriously for quite a long time" (Nelson and Nelson 1996:44).

Scenario 3

"Almost everybody, at some point in life, will avoid uncomfortable truths, 'edit' their own memories, mislead others, and even sometimes tell out-and-out falsehoods. And almost everybody feels uncomfortable about lying repeatedly. As Barbara wrestles with this problem, she has put herself in her dad's shoes and acknowledged that she would feel very uncomfortable if it turned out that someone was lying to her. Even so, she also knows from experience that the price of avoiding a lie can sometimes be just as high as the price of telling one.

"While people will agree that one ought to tell the truth whenever possible, it's not so easy to say precisely why that's so. To understand better whether and when it's morally okay to break the rule against lying, it's necessary to figure out just what's at stake in telling the truth or failing to do so" (Nelson and Nelson 1996:25).

Scenario 4

"I use a wheelchair because I was paralyzed by polio 40 years ago. One of my first trips out of the hospital back then was to a supermarket. I remember I was rolling down an aisle when a kid saw me. He stopped dead in his tracks and pointed. 'Mommy,' he said in a loud voice.....and in a loud voice, 'Mommy, look at the broken man'" (Gallagher 1992).

Scenario 5

"Lauren M. Cook had been participating in reenactments of famous Civil War battles for two years, and she took the hobby seriously. She spent thousands of dollars buying Civil War period clothing. She bound her breasts under her uniform so no one would know she was a woman. She even tried to adopt male mannerisms to aid her disguise. 'I would always squint,' she said. 'Women's eyes are larger than men's, so they really give you away'" (Marcus 2002).

Practice Test: Multiple-Choice Questions

1. The Congo is emphasized in Chapter 5 ("Social Interaction) because
    a. HIV originated in the Congo.
    b. HIV "traveled" from Europe to the Congo.
    c. a blood sample frozen in 1959 and stored in a Congo blood bank provides evidence that HIV existed before the 1980s.
    d. HIV traveled from the Congo to Cuba to the United States.

2. _____ wrote *The Division of Labor in Society*.
    a. Karl Marx
    b. Max Weber
    c. Emile Durkheim
    d. C. Wright Mills

3. King Leopold II claimed the Congo as his private property. His reign over the land has been described as
    a. important to the economic progress of the Belgian Congo.
    b. the vilest scramble for loot that ever disfigured the history of human conscience and geographical location.
    c. putting the Congo and its people in the modern world.
    d. the event that led to the formation of a democratic form of government.

4. Emile Durkheim used the general term _____ to describe the ties that bind people to one another in a society.
    a. social interaction
    b. solidarity
    c. context
    d. content

5. Durkheim wrote that a person's "first duty is to resemble everybody else—to not have anything personal about one's core beliefs and actions." Durkheim was writing about
    a. the division of labor.
    b. specialization.
    c. mechanical solidarity.
    d. organic solidarity.

6. According to Durkheim, the vulnerability of societies _____ as the division of labor becomes more complex and specialized.
    a. decreases
    b. remains the same
    c. increases
    d. increases but eventually decreases

7. Which one of the following disruptions to the division of labor occurs when workers are so isolated that few people grasp the workings and consequences of the overall enterprise?
   a. job specialization
   b. industrial and commercial crises
   c. inefficient management of worker talents
   d. forced division of labor

8. After achieving independence in 1960, the Belgian Congo became
   a. the Democratic Republic of Congo.
   b. the Congo.
   c. Zaire.
   d. the Congo Republic.

9. In which of the following countries was the medical condition HIV/AIDS was *first* noticed in the late 1980s?
   a. the United States
   b. People's Republic of China
   c. Belgium
   d. Haiti

10. When physicians knowingly perform unnecessary surgery, overmedicate, engage in sexual relations with patients, or break confidentiality, they have
    a. failed to meet role obligations.
    b. experienced role strain.
    c. no right to demand patient cooperation.
    d. achieved role conflict.

11. Which one of the following generalizations is <u>least</u> characteristic of Western-style medicine?
    a. Western physicians rely heavily on technological tools to diagnose and treat illness.
    b. The major objective of the patient-physician interaction is to determine the exact physiological problem.
    c. Tremendous effort is devoted to finding a technological solution to illness.
    d. When diagnosing illness, Western physicians attach considerable importance to the patient's social relationships and psychological distress.

12. The _____ model corresponds to the perspective in which social interaction is viewed as though it is taking place in a theater.
    a. dramaturgical
    b. historical
    c. cultural strain
    d. division of labor

13. On the first day of class, Professor Smith always wears a tie to convey that he is serious about his job. On the other hand, he gives out his home number as a way of letting students know he is approachable. Professor Smith is engaged in
    a. backstage behavior.
    b. impression management.
    c. role strain.
    d. role conflict.

14. _____ is a useful concept for understanding the dilemma that sexual partners face when one partner suggests using a condom as a precautionary condition of sexual intercourse.
    a. Role
    b. Impression management
    c. Social structure
    d. Role strain

15. From a sociological point of view, restaurant kitchen employees who eat food from customers' plates are engaging in _____ behavior.
    a. frontstage
    b. backstage
    c. upfront
    d. negligent

16. Attributing cause to _____ factors functions to reduce uncertainty about the source and spread of disease.
    a. dispositional
    b. situational
    c. backstage
    d. contextual

17. Isiah argues that HIV-infected people earned their disease as a penalty for their perverse, indulgent, and illegal behaviors. Isiah is attributing HIV to
    a. dispositional factors.
    b. situational factors.
    c. role conflict.
    d. role strain.

18. In sociological terms, a(n) _____ is a person or a group that is assigned blame for conditions that threaten a community's sense of well-being or shake the foundations of a trusted institution.
    a. attribution
    b. target
    c. scapegoat
    d. disposed person

19. We now know that 50 percent of _____ were HIV-infected before the first case of AIDS appeared in this group.
     a. homosexuals
     b. hemophiliacs
     c. Haitians
     b. drug users

20. The U.S. supplies about _____ percent of the world's blood and blood products.
     a. 10
     b. 15
     c. 25
     d. 60

True/False Questions

1. T F   HIV's origin cannot be understood apart from European colonial rule of Africa.

2. T F   The purpose of the West Africa Conference in 1885 was to divide Africa among competing European powers.

3. T F   The Mbuti pygmies share a forest-oriented value system.

4. T F   Sociologists define social status as rank of prestige.

5. T F   In the United States, 40 percent of airline pilots are female.

6. T F   Upon learning their health status, nine of ten HIV-infected people do not donate blood.

7. T F   Twenty percent of blood donors claim that they would have answered screening questions differently in a more private setting.

8. T F   When evaluating the causes of their own failures, people tend to favor situational factors.

9. T F   Attributions about who should have AIDS affects the way AIDS is defined and diagnosed

10. T F   We simply do not know how many people are infected with HIV in the United States or worldwide.

- **Union of Concerned Scientists**
  http://www.ucsusa.org/
  The Union of Concerned Scientists, an independent nonprofit alliance of more than 100,000 concerned citizens and scientists seeks to "build public awareness of global environmental issues" including antibiotic resistance, biodiversity, fuel-efficient vehicles, genetically engineered food, global warming, health and environment, missile defense, nuclear power, and renewable energy.

- **Democratic Republic of the Congo (Zaire) Page**
  http://www.sas.upenn.edu/African_Studies/Country_Specific/Zaire.html
  The African Studies Center at the University of Pennsylvania posts country-specific pages on sub-Saharan African countries including the Democratic Republic of the Congo (formerly Zaire).

Applied Research

Read *The New York Times, Los Angeles Times*, or some other newspaper with a national reputation for one week. Clip out or print articles about disruptions in the division of labor as defined by Durkheim. Do the articles give insight about how disruptions break down people's ability to connect with one another in meaningful ways through their labor?

Background Notes: Democratic Republic of the Congo

Size: about the size of the U.S. east of the Mississippi
Capital: Kinshasa
Population (2005): 58 million
Ethnic groups: More than 200 African ethnic groups; the Luba, Kongo, and Anamongo are some of the larger groupings of tribes
Religions: Roman Catholic 50%, Protestant 20%, other syncretic sects and traditional beliefs 10%, Kimbanguist 10%, Muslim 10%
Infant mortality rate: 94.7/1,000 births
Life expectancy at birth: female 49 years

The Democratic Republic of the Congo (D.R.C.) includes the greater part of the Congo River basin, which covers an area of almost 1 million square kilometers (400,000 sq. mi.). The country's only outlet to the Atlantic Ocean is a narrow strip of land on the north bank of the Congo River.

The area known as the Democratic Republic of the Congo was populated as early as 10,000 years ago and settled in the 7th and 8th centuries A.D. by Bantus from present-day Nigeria. Discovered in 1482 by Portuguese navigator Diego Cao and later explored by English journalist Henry Morton Stanley, the area was officially colonized as the Congo Free State in 1885 as a personal possession of Belgian King Leopold II. In 1907, administration shifted to the Belgian Government, which renamed the country the Belgian Congo. Following a series of riots and unrest, the Belgian Congo was granted its independence on June 30, 1960. Parliamentary elections in 1960 produced Patrice Lumumba as prime minister and Joseph Kasavubu as president of the renamed Democratic Republic of the Congo.

Within the first year of independence, several events destabilized the country: the army mutinied; the governor of Katanga province attempted secession; a UN peacekeeping force was called in to restore order; Prime Minister Lumumba died under mysterious circumstances; and Col. Joseph Désiré Mobutu (later Mobutu Sese Seko) took over the government and ceded it again to President Kasavubu.

Unrest and rebellion plagued the government until 1965, when Lieutenant General Mobutu, by then commander-in-chief of the national army, again seized control of the country and declared himself president for 5 years. Mobutu quickly centralized power into his own hands and was elected, without opposition, as president in 1970. Embarking on a campaign of cultural awareness, Mobutu renamed the country the Republic of Zaire and required citizens to adopt African names. Relative peace and stability prevailed until 1977 and 1978 when Katangan rebels staged in Angola launched a series of invasions into the Katanga region. The rebels were driven out with the aid of Belgian paratroopers.

During the 1980s, Mobutu continued to enforce his one-party system of rule. Although Mobutu successfully maintained control during this period, opposition parties, most notably the Union pour la Democratie et le Progres Social (UDPS), were active. Mobutu's attempts to quell these groups drew significant international criticism.

As the Cold War came to a close, internal and external pressures on Mobutu increased. In late 1989 and early 1990, Mobutu was weakened by a series of domestic protests, by heightened international criticism of his regime's human rights practices, and by a faltering economy. In April 1990, Mobutu agreed to the principle of a multi-party system with elections and a constitution. As details of a reform package were delayed, soldiers in September 1991 began looting Kinshasa to protest their unpaid wages. Two thousand French and Belgian troops, some of whom were flown in on U.S. Air Force planes, arrived to evacuate the 20,000 endangered foreign nationals in Kinshasa.

In 1992, after previous similar attempts, the long-promised Sovereign National Conference was staged, encompassing more than 2,000 representatives from various political parties. The conference gave itself a legislative mandate and elected Archbishop Laurent Monsengwo as its chairman, along with Etienne Tshisekedi, leader of the UDPS, as prime minister. By the end of the year, Mobutu had created a rival government with its own prime minister. The ensuing stalemate produced a compromise merger of the two governments into the High Council of Republic-Parliament of Transition (HCR-PT) in 1994, with Mobutu as head of state and Kengo Wa Dondo as prime minister. Although presidential and legislative elections were repeatedly scheduled over the next two years, they never took place.

By 1996, the war and genocide in neighboring Rwanda had spilled over to Zaire. Rwandan Hutu militia forces (Interahamwe) that fled Rwanda following the ascension of a Tutsi-led government were using Hutu refugee camps in eastern Zaire as bases for incursions against Rwanda.

In October 1996, Rwandan troops (RPA) entered Zaire, simultaneously with the formation of an armed coalition led by Laurent-Desire Kabila known as the Alliance des Forces Democratiques pour la Liberation du Congo-Zaire (AFDL). With the goal of forcibly ousting Mobutu, the AFDL, supported by Rwanda and Uganda, began a military campaign toward Kinshasa. Following failed peace talks between Mobutu and Kabila in May 1997, Mobutu left the country, and Kabila marched into Kinshasa on May 17, 1997. Kabila declared himself president, consolidated power around himself and the AFDL, and renamed the country the Democratic Republic of Congo (DRC). Kabila's Army Chief and the Secretary General of the AFDL were Rwandan, and RPA units continued to operate tangentially with the DRC's military, which was renamed the Forces Armees Congolaises (FAC).

Over the next year, relations between Kabila and his foreign backers deteriorated. In July 1998, Kabila ordered all foreign troops to leave the DRC. Most refused to leave. On August 2, fighting erupted throughout the DRC as Rwandan troops in the DRC "mutinied," and fresh Rwandan and Ugandan troops entered the DRC. Two days later, Rwandan troops flew to Bas-Congo, with the intention of marching on Kinshasa, ousting Laurent Kabila, and replacing him with the newly formed Rwandan-backed rebel group called the Rassemblement Congolais pour la Democratie (RCD). The Rwandan campaign was thwarted at the last minute when Angolan, Zimbabwean, and Namibian troops intervened on behalf of the DRC Government. The Rwandans and the RCD withdrew to eastern DRC, where they established de facto control over portions of the eastern DRC and continued to fight the Congolese Army and its foreign allies.

In February 1999, Uganda backed the formation of a rebel group called the Mouvement pour la Liberation du Congo (MLC), which drew support from among ex-Mobutuists and ex-FAZ soldiers in the Equateur province (Mobutu's home province). Together, Uganda and the MLC established control over the northern third of the DRC.

At this stage, the DRC was divided de facto into three segments, and the parties controlling each segment had reached military deadlock. In July 1999, a cease-fire was proposed in Lusaka, Zambia, which all parties signed by the end of August. The Lusaka Accord called for a cease-fire, the deployment of a UN peacekeeping operation, MONUC, the withdrawal of foreign troops, and the launching of an "Inter-Congolese Dialogue" to form a transitional government leading to elections. The parties to the Lusaka Accord failed to fully implement its provisions in 1999 and 2000. Laurent Kabila drew increasing international criticism for blocking full deployment of UN troops, hindering progress toward an Inter-Congolese Dialogue, and suppressing internal political activity.

On January 16, 2001, Laurent Kabila was assassinated. He was succeeded by his son Joseph Kabila. Joseph Kabila reversed many of his father's negative policies; over the next year, MONUC deployed throughout the country, and the Inter-Congolese Dialogue proceeded. By the end of 2002, all Angolan, Namibian, and Zimbabwean troops had withdrawn from the DRC. Following DRC-Rwanda talks in South Africa that culminated in the Pretoria Accord in July 2002, Rwandan troops officially withdrew from the DRC in October 2002, although there were continued, unconfirmed reports that Rwandan soldiers and military advisers remained integrated with RCD/G forces in the eastern DRC. Ugandan troops officially withdrew from the DRC in May 2003.

In October 2001, the Inter-Congolese Dialogue began in Addis Ababa under the auspices of Facilitator Ketumile Masire (former president of Botswana). The initial meetings made little progress and were adjourned. On February 25, 2002, the dialogue was reconvened in South Africa. It included representatives from the government, rebel groups, political opposition, civil society, and Mai-Mai (Congolese local defense militias). The talks ended inconclusively on April 19, 2002, when the government and the MLC brokered an agreement that was signed by the majority of delegates at the dialogue but left out the RCD/G and opposition UDPS party, among others.

This partial agreement was never implemented, and negotiations resumed in South Africa in October 2002. This time, the talks led to an all-inclusive power sharing agreement, which was signed by delegates.

Sparsely populated in relation to its area, the Democratic Republic of the Congo is home to a vast potential of natural resources and mineral wealth. Nevertheless, the DRC is one of the poorest countries in the world, with per capita annual income of about $98 in 2003. This is the result of years of mismanagement.

Agriculture is the mainstay of the Congolese economy, accounting for 56.3 percent of GDP in 2002. The main cash crops include coffee, palm oil, rubber, cotton, sugar, tea, and cocoa. Food crops include cassava, plantains, maize, groundnuts, and rice. Industry in the DRC, especially the mining sector, is underdeveloped relative to its potential. In 2002, industry accounted for only 18.8 percent of GDP, with only 3.9 percent attributing to manufacturing. Services reached 24.9 percent of GDP. The Congo was the world's fourth-largest producer of industrial diamonds during the 1980s, and diamonds continue to dominate exports, accounting for over half of exports ($642 million) in 2003. The Congo's main copper and cobalt interests are dominated by Gecamines, the state-owned mining giant. Gecamines production has been severely affected by corruption, civil unrest, world market trends, and failure to reinvest.

Answers

Concept Applications
1. Role strain
2. Front stage, Impression management
3. Impression management
4. Dispositional traits
5. Impression management

Multiple-Choice		**True/False**
1. c page 117	11. d page 129	1. T page 117
2. c page 118	12. a page 130	2. T page 118
3. b page 119	13. b page 130	3. T page 121
4. b page 121	14. b page 131	4. F page 125
5. c page 121	15. b page 131	5. F page 127
6. c page 122	16. a page 133	6. F page 133
7. a page 122	17. a page 133	7. T page 133
8. c page 124	18. c page 134	8. T page 133
9. b page 126	19. b page 136	9. T page 135
10. a page 128	20. d page 136	10. T page 136

hapter 6

Formal Organizations

Study Questions

1. Why is the McDonald's corporation the focus of a chapter on formal organizations?

2. What is a formal organization?

3. Distinguish between primary and secondary groups? Name and define the three kinds of secondary groups that exist.

4. What is a bureaucracy? Is McDonald's a bureaucracy? Explain.

5. How is studying a bureaucracy as an ideal type useful?

6. Distinguish between formal and informal dimensions of organizations.

7. What is value-rational thought? Why was Max Weber especially concerned with value-rational action?

8. Define rationalization. How does the example of factory farms relate to rationalization? What are the positive and negative outcomes of rationalization?

9. What is McDonaldization of society?

10. Explain the iron cage of rationality.

11. How do organizations, such as McDonald's, reach beyond local markets to regional, national, and global markets?

12. Define multinational corporation. In what ways are multinational corporations engines of progress? In what ways are they engines of destruction?

13. How do sociologists demonstrate the size and, by extension, power of global corporations? Give examples.

14. What are externality costs?

15. How does the case of Margie Eugene Richard, who took on Shell Chemical, illustrate the power of informed consumers?

16. What is trained incapacity? Give an example from Shoshana Zuboff's *In the Age of the Smart Machine* of a work environment that promotes trained incapacity. Contrast that work environment with one that promotes empowering behavior.

17. How do organizations use statistical measures of performance? What are some of the problems that can accompany such measures? Give examples of statistical measures of performance at McDonald's.

18. What is expert power? How can expert power be problematic?

19. Define oligarchy. Why does oligarchy seem to be an inevitable feature of large organizations?

20. How did Karl Marx define alienation? What are the four levels of alienation? Give examples of each type.

Concept Applications

Consider the concepts listed below. Match one or more of the concepts with each scenario. Explain your choices.

    a. Automate
    b. Externality costs
    c. Informal dimensions of organizations
    d. Multinational corporation
    e. Value-rational

Scenario 1

"Is IBM Japan an American or a Japanese company? Its workforce of 20,000 is Japanese, but its equity holders are American. Even so, over the past decade, IBM Japan has provided, on average, three times more tax revenue to the Japanese government than has Fujitsu. What is its nationality? Or what about Honda's operation in Ohio? Or Texas Instruments' memory-chip activities in Japan? Are they 'American' products? If so, what about the cellular phones sold in Tokyo that contain components made in the United States by American workers who are employed by the U.S. division of a Japanese company? Sony has facilities in Dotham, Alabama, from which it sends audio tapes and video tapes to Europe. What is the nationality of these products or the operation that makes them?" (Ohmae 1990:10).

Scenario 2

"A number of employees (5%) respond to perceived injustices by not performing their required tasks. One incident involved a male stockroom worker at a retail store who claimed he was paid less than others in similar positions. After an unsuccessful attempt to discuss the matter with his supervisor, the worker decided to deal with the conflict in his own way: 'I didn't really want to quit, so I goofed off a lot. I didn't do anything unless I was specifically asked to. When working at night, I would listen to music for hours and do nothing.... If I was goofing off and saw the manager, I would act as if I was really doing something'" (Tucker 1993:37).

Scenario 3

"Scientifically, the atomic bomb was an advance into unknown territory, but militarily, it was simply a more cost-effective way of attaining a goal that was already a central part of strategy: a means of producing the results achieved at Hamburg and Dresden cheaply and reliably every time the weapon was used [for example, a quarter million bombs were used to destroy the city of Dresden]. Even at the time, the $2 billion cost of the Manhattan Project was dwarfed by the cost of trying to destroy cities the hard way, using conventional bombs" (Dyer 1985:96).

Scenario 4

"The same kind of computer technology that enables employers to keep track of workers' backgrounds also makes it possible for them to quantify and monitor work performance. Anyone who works on a video display terminal, electronic telephone console, or other computer-based equipment, including laser scanner cash registers, is subject to constant monitoring.

"Although the stated aim of monitoring workers is to improve productivity and service, the effect can be to turn checkstands into pressure cookers. 'Computers are wonderful for many things,' says Beverly Crownover, president of Local 1532 of United Food and Commercial Workers in Santa Rosa, California. 'But when they're used to monitor how many items a cashier scans per minute, it's like a whip. There's incredible pressure on workers'" (UFCW Action 1993:135).

Scenario 5

"The cost of stress to the American workplace has been estimated at between $150 billion and $180 billion a year. Stress-related illness accounts for millions of lost working days each year, and the number is rising. One study found that in 1980, no occupational disease claims were related to stress; in 1990, 10 percent of them were. A 1993 study by Commerce Clearing House reports that unscheduled absences can cost U.S. employers more than $500 per employee per year. Experts believe that stress accounts for 12 percent of all workers' compensation claims" (Wright and Smye 1996:7).

Practice Test: Multiple-Choice Questions

1. From a sociological point of view, a formal organization is
    a. a legally recognized group of people.
    b. a coordinating mechanism created by people to achieve stated objectives.
    c. the building in which people meet.
    d. a money making enterprise.

2. Which of the following is the <u>best</u> example of a formal organization?
    a. the class of '95
    b. shoppers in a mall
    c. Wal-Mart
    d. the country of India

3. If relationships between people are limited to a specific activity and setting, the people are part of a(n)
    a. primary group.
    b. secondary group.
    c. outgroup.
    d. ingroup.

4. _____ draw in people seeking material gain in the form of pay, health benefits, or a new status.
    a. Voluntary organizations
    b. Coercive organizations
    c. Utilitarian organizations
    d. Bureaucracies

5. According to standard operating procedures, every customer at McDonald's is greeted with the words, "Welcome to McDonald's. May I take your order?" The practice corresponds with which characteristic of a bureaucracy?
    a. a clean-cut division of labor
    b. positions filled on the basis of qualification
    c. personnel treat "clients" as cases
    d. authority belongs to the position

6. In a British-based McDonald's court case, employees testified that they routinely witnessed managers and employees watering down soft drinks and failing to throw away food that had been dropped on the floor. These observations relate to
    a. oligarchy.
    b. the ideal type.
    c. the informal dimensions of organizations.
    d. the formal dimension of organizations.

7. Theoretically, in a bureaucracy,
    a. authority belongs to the person.
    b. positions are filled on the basis of connections.
    c. authority resides in the personalities of people holding important positions.
    d. personnel treat clients as cases and without emotion.

8. From a value-rational point of view, nature is something to be
    a. conserved.
    b. respected.
    c. held in awe.
    d. used to make a profit.

9. A hamburger purchased in a Wyoming Wendy's and a hamburger purchased in a German Wendy's has the same appearance and taste. This phenomenon represents which feature of McDonaldization?
    a. efficiency
    b. predictability
    c. control
    d. quantification

10. Outside of North America, what region of the world is *least likely* to have a McDonald's?
    a. Asia
    b. South America
    c. Europe
    d. Western and Central Africa

11. Which one of the following is an example of an externality cost?
    a. the cost of labor
    b. the cost of materials to produce a product
    c. the cost of operating a manufacturing plant
    d. the cost of restoring contaminated and barren land

12. A worker says, "Sometimes, I am amazed when I realize that we stare at the computer screen even when it has gone down." This comment suggests that in that organization, computers are used as
    a. an automating tool.
    b. an informating tool.
    c. a coordinating mechanism.
    d. a technological resource.

13. If occupational safety is measured by the number of accidents that occur on the job, then the _____ industry has one of the lowest accident rates of all industries.
    a. construction
    b. logging
    c. chemical
    b. trucking

14. June works as a cashier. Her productivity is judged according to the number of items passed over a scanner per hour. She is being rated according to
    a. statistical measures of performance.
    b. trained incapacity.
    c. informal policies.
    d. an oligarchy.

15. McDonald's employs about 500,000 people in its corporate offices. It is impossible for that many people to come together to discuss issues that affect daily operations. This shortcoming reflects the principles of
    a. oligarchy.
    b. bureaucracy.
    c. expert power.
    d. McDonaldization.

16. The danger of oligarchy is that those who make decisions may
    a. run the organization as a bureaucracy.
    b. not have the necessary background to understand the full implications of their decisions.
    c. rely on informal mechanisms to get things done.
    d. suffer from disenchantment of the world.

17. Fertilizers, herbicides, pesticides, and chemically treated seeds give people control over nature because they eliminate the need to fight weeds, and they prevent pests from destroying crops. Yet heavy reliance on chemical technologies causes the soil to erode and become less productive. This dilemma represents a case of
    a. alienation.
    b. trained incapacity.
    c. oligarchy.
    d. optimum technology.

18. Employers that specify exactly how workers should look, behave, and speak are contributing to
    a. informal dimensions of control.
    b. a safe work environment.
    c. informating the workplace.
    d. alienation from self.

19. Workers are alienated from _____ because home and work environments are separate.
    a. the process of production
    b. the product
    c. the family
    d. the self

20. Because employers own the factory buildings, the tools, the machines, and the labor of workers, the workers are alienated from
    a. the process of production.
    b. the product.
    c. the family.
    d. the self.

True/False Questions

1. T F   Formal organizations have a life that extends beyond the people that comprise them.

2. T F   Secondary groups can range in size from small to extremely large.

3. T F   Organizational problems occur when employees fail to follow official policies *or* follow them too rigidly.

4. T F   Actual behavior in organizations departs from the ideal.

5. T F   Multinational corporations plan, produce, and sell on a national scale.

6. T F   Multinational corporations are headquartered disproportionately in the United States, Japan, and Western Europe.

7. T F   France is one of the five countries in the world with GNP that exceeds the annual combined revenues of the top 10 global corporations.

8. T F   To informate means to use the computer as a source of surveillance.

9. T F   Karl Marx is the theorist associated with the concept *alienation*.

10. T F   Oligarchy means rule by the many.

Internet Resources

- **The Global Web100**
  http://metamoney.com/globalListIndex.html
  Global Web100 offers a list of links to the 100 largest non-U.S. based global corporations with webpages on the internet. See http://metamoney.com/usListIndex.html for the 100 largest U.S.-based corporations.

- **Multinational Monitor**
  http://www.essential.org/monitor/monitor.html
  Multinational Monitor is "a monthly magazine devoted primarily to examining the activities of multinational companies." Each issue features a special topic. Examples include "Medicine and Market" "The Case Against GE," "Corporations and the U.S. Poor," and "The Ten Worst Corporations of 2003."

Applied Research

Select one of the world's top 10 largest global corporations listed in your textbook. Research how the corporation you selected reaches around the globe through joint ventures, supply deals for parts and components, assembly operations, and marketing and distribution offices.

Answers

Concept Applications
1. Multinational corporation
2. Informal dimensions of organizations
3. Value-rational
4. Automate
5. Externality costs

Multiple-Choice						
1.	b	page 144	11.	d	page 155
2.	c	page 144	12.	a	page 158
3.	b	page 144	13.	c	page 159
4.	c	page 145	14.	a	page 158
5.	c	page 146	15.	a	page 160
6.	c	page 147	16.	b	page 160
7.	d	page 146	17.	a	page 161
8.	d	page 148	18.	d	page 163
9.	b	page 149	19.	c	page 162
10.	d	page 152	20.	a	page 161

True/False		
1.	T	page 143
2.	T	page 144
3.	T	page 147
4.	T	page 146
5.	F	page 151
6.	T	page 152
7.	F	page 153
8.	F	page 158
9.	T	page 161
10.	F	page 159

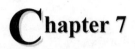hapter 7

Deviance, Conformity, and Social Control

Study Questions

1. Why focus on China in conjunction with concepts of deviance, conformity, and social control?

2. What is deviance? How is it related to conformity and social control?

3. Is it possible to generate a list of deviant behaviors? Why or why not?

4. Distinguish between folkways and mores. Give examples of each concept.

5. What important cultural lessons are incorporated into the daily activities of Chinese and American preschoolers? How do parents from each culture react to the other's system?

6. What are the major mechanisms of social control? Give examples of each.

7. According to Durkheim, why is crime a "normal" and necessary phenomenon?

8. What are the major assumptions that guide labeling theory? How do these assumptions relate to the following categories: conformists, pure deviant, secret deviant, and falsely accused?

9. Under which circumstances are people most likely to be falsely accused of a crime?

10. What are witch hunts? Why do they occur? Give an example of witch hunt.

11. Define white-collar crime and corporate crime. Why are white-collar and corporate criminals less likely to be caught than so-called common criminals?

12. In Milgram's classic experiment *Obedience to Authority*, why did a significant number of volunteers come to accept an authority's definition of deviance and administer shocks even though these shocks caused obvious harm to confederates?

13. Who are claims makers? What factors determine a claim maker's success?

14. Describe the constructionist approach to analyzing claims makers and claims making activities.

15. What is structural strain? What are the sources of structural strain in the United States?

16. What are the responses to structural strain?

17. Identify one source of structural strain in China. Use Merton's typology of responses to consider how people in China respond to this strain.

18. Summarize the major assumptions underlying the theory of differential association. How does this assumption relate to the mechanisms of social control in China?

19. What larger historical and geographical factors shape China's one-child policy?

Concept Applications

Consider the concepts listed below. Match one of more of the concepts with each scenario. Explain your choices.

a. Claims makers
b. Falsely accused
c. Mores
d. Secret deviants
e. White-collar crimes

Scenario 1

"The Tobacco Institute was founded in 1958, even before the first Surgeon General's report on the health risks of smoking, to represent the interests of tobacco companies to lawmakers. Once financed by a dozen companies, it now works for only five—Philip Morris, R. J. Reynolds, Lorillard, Liggett, and American Brands—but its twofold mission remains the same: to persuade federal, state, and local authorities to lay off and to sell the virtues of the industry to the American public. A staff of lobbyists handles the first task and Ms. Dawson, at 32, the second. The job description is fairly typical for a trade organization—to develop and articulate the industry position on any given issue, then make sure the message reaches the public. But this is no typical industry" (Janofsky 1994:8F).

Scenario 2

"Boesky told the government about his insider trading activities, not only with me but with at least one other well-known investment banker. Beyond that, he detailed various schemes, concocted with those in the highest circles of power, to circumvent SEC regulations and tax laws. Said Carroll, 'He has played fast and loose with the rules that govern our markets, with the effect of manipulating the outcome of financial transactions measured in the hundreds of millions of dollars'" (Levine and Hoffer 1991:346).

Scenario 3

"The small-time criminals are everywhere. Maybe they're sneaking into more than one theater in the local Cineplex or grabbing a handful of yogurt peanuts from the grocery store bin and eating all the evidence before getting to the check-out stand or making personal long-distance calls from work" (Tomashoff 1993:E1).

Scenario 4

"Death sentences for people who later prove to be innocent are less unusual than is commonly supposed. Just in the last five months, four once-condemned prisoners have been released after spending years on death row. Two of them, in Alabama and Texas, turned out to have been convicted on fabricated evidence and perjured testimony; the third, in Texas, was convicted because of evidence that was withheld; the fourth, in Maryland, was exonerated by DNA analysis, a technology that was unavailable at the time of his trial" (*The New Yorker* 1993:14).

Scenario 5

"Can a court force an unwilling person to give up part of his or her body (e.g., bone marrow, a kidney) to a relative who needs that body part to survive? That was the question recently brought before the court of common pleas in Allegheny County, Pennsylvania. The common law has consistently held that one human being is under no legal obligation to give aid or take action to save another human being or to rescue one. The court said that such a rule, although revolting in a moral sense, is founded upon the very essence of a free society, and while other societies may view things differently, our society has as its first principle respect for the individual—and society and government exist to protect that individual from being invaded and hurt by another" (Chayet 1983).

Practice Test: Multiple-Choice Questions

1. The sociological contribution to understanding deviant behavior is the emphasis on
    a. the individual as a deviant.
    b. unchanging and universal definitions of deviance.
    c. the context under which deviant behavior occurs.
    d. conformity.

2. _____ is any behavior or appearance that follows and maintains the standards of a group.
   a. Deviance
   b. Conformity
   c. Social control
   d. Power

3. Which one of the following characteristics applies to the concept of mores?
   a. essential to the well-being of a group
   b. one of many ways to do things
   c. routine matters
   d. details of life

4. A preschool teacher comments, "Why do small children in the United States go to the bathroom separately? It is much easier to have everyone go at the same time." This comment is most likely coming from a teacher working in
   a. China.
   b. Taiwan.
   c. Japan.
   d. South Korea.

5. Six-year-old Martha picks up her toys and puts them away. Her father smiles and pats her on the back. The smile and pat represent a _____ sanction.
   a. positive formal
   b. negative formal
   c. negative informal
   d. positive informal

6. A Harvard Law School study identified _____ as the country with the most extensive internet censorship in the world.
   a. South Korea
   b. the United States
   c. China
   d. Romania

7. The U.S. Department of Justice asked meter readers, cable installers, and telephone repair people to report suspicious activities they might notice while serving customers. This qualifies as a form of
   a. censorship.
   b. surveillance.
   c. conformity.
   d. deviance.

8. Durkheim's theory of deviance (crime) is written from a _____ perspective.
   a. functionalist
   b. conflict
   c. symbolic interactionist
   d. social action

9. Durkheim's theory of deviance (crime) does not address which one of the following questions?
   a. Why is deviance present in every society?
   b. How can almost any behavior qualify as deviant?
   c. Who decides that a particular activity or appearance is deviant?
   d. How is deviance functional for society?

10. The "Great Leap Forward" was Mao's plan, which
    a. pushed the country into the 21st century.
    b. was supported by the United States.
    c. created economic surpluses and generated prosperity.
    d. failed miserably.

11. A U.S. Bureau of Justice survey of crime victims documented that almost 58 percent of crime victims do not report the crime to police. This suggests that there are large numbers of _____ in U.S. society.
    a. conformists
    b. pure deviants
    c. secret deviants
    d. falsely accused

12. A campaign to identify, investigate, and correct behavior that is believed to undermine a group or a country is known as
    a. a moral event.
    b. ethnic cleansing.
    c. a witch hunt.
    d. target practice.

13. Under which of the following circumstances is a person least likely to be falsely accused of committing a crime?
    a. times of economic crisis
    b. during a health crisis (or epidemic)
    c. times of rapid economic growth
    d. when an important institution is threatened

14. Former Governor George Ryan argued that "our capital system is haunted by the demon of error—error in determining guilt and error in determining who among the guilty deserve to die." Ryan focuses our attention on which one of the following parties?
    a. the criminal
    b. rule makers and rule enforcers
    c. the guilty
    d. the innocent

15. _____ is the researcher who conducted the classic study *Obedience to Authority*.
    a. Stanley Milgram
    b. Emile Durkheim
    c. Howard Becker
    d. Erving Goffman

16. When constructionists study the process by which a group or behavior is defined as a problem to society, they focus on
    a. the valued goals and the means to achieve those goals.
    b. the rule breaker, rule maker, and rule enforcers.
    c. responses to structural strain.
    d. who makes the claims, whose claims are heard, and how audiences respond.

17. The Chinese government issued the report "The Human Rights Record of the United States because
    a. it believes that the United States is a model with regard to human rights.
    b. the U.S. asked for an independent evaluation of its human rights record.
    c. it wanted to show that its human rights record is better than the U.S record.
    d. the U.S. issues Country Reports on Human Rights Practices each year for 190 countries but does not critique its own record.

18. The students who participated in the 1989 demonstration in Tiananmen Square were accused of "bourgeois liberalization." This means they were guilty of
    a. following capitalist principles.
    b. draft evasion.
    c. the wanton expression of individual freedom.
    d. slandering Chairman Mao.

19. Roughly, one of every _____ people alive in the world lives in the People's Republic of China.
    a. five
    b. two
    c. ten
    d. twenty

20. The *habitable* land area of China is approximately
    a. the size of the United States.
    b. the size of Europe, North America, and South America combined.
    c. half the size of the United States.
    d. the size of the African continent.

True/False Questions

1. T F   The sociological contribution to deviance is that it focuses on the deviant individual.

2. T F   The Cultural Revolution occurred between 1966 and 1976.

3. T F   We acknowledge the previous legality of cocaine in the United States whenever we ask for a "coke".

4. T F   Conceptions of what is deviant vary across time and place.

5. T F   Ridicule is a formal sanction.

6. T F   The United States has the highest incarceration rate in the world.

7. T F   According to Emile Durkheim, deviance will be present even in a community of saints.

8. T F   In the U.S., about 90 percent of crime victims report the crime to police.

9. T F   Stanley Milgram's study *Obedience to Authority* is relevant to understanding the Cultural Revolution and Abu Ghraib.

10. T F   The U.S. State Department classifies China's human rights record as poor.

Internet Resources

*   **Bureau of Justice Statistics**
    http://www.ojp.usdoj.gov/bjs
    The Bureau of Justice Statistics "collects, analyzes, publishes, and disseminates information on crime, criminal offenders, victims of crime, and the operation of justice systems at all levels of government." According to the Bureau the data it collects is pivotal to "combating crime and ensuring that justice is both efficient and evenhanded."

- **The Innocence Project**
  http://www.innocenceproject.org/
  A non-profit organization that has exonerated 154 convicted prisoners on death row since 1992. The organization fights for "poor and forgotten" clients by using biological DNA evidence to prove innocence.

- **The China Daily**
  http://www.chinadaily.com.cn/english/home/index.html
  This daily China news site serves as "an online bridge between China and the rest of the world" featuring "news, information, services and education to four million viewers a day."

Applied Research

The internet has more than 20,000 crime-stopper sites. The Peel Regional Police website defines Crime Stoppers as a "nonprofit organization which rallies the community, the news media and the police in a collective campaign against crime. Crime Stoppers' mandate is to fight crime." Use one of the internet search engines (i.e., Infoseek, Yahoo, Excite) to find and review a small sample of crime stopper websites. After viewing these websites, consider how Durkheim would react to and explain their existence.

Background Notes: Peoples Republic of China

Size: Slightly smaller than the U.S.
Capital: Bejing
Population (2005): 1.3 billion
Ethnic groups: *Han Chinese*—91.9%; Zhuang, Manchu, Hui, Miao, Uygur, Yi, Mongolian, Tibetan, Buyi, Korean, and *other nationalities*—8.1%
Religions: Officially atheist; Taoism, Buddhism, Islam, Christianity
Infant mortality rate: 23.1/1,000 births
Life expectancy at birth: female 74.5 years; males 70.9 years

Religion plays a significant part in the life of many Chinese. Buddhism is most widely practiced, with an estimated 100 million adherents. Traditional Taoism also is practiced. Official figures indicate there are 20 million Muslims, 5 million Catholics, and 15 million Protestants; unofficial estimates are much higher.

While the Chinese constitution affirms religious toleration, the Chinese government places restrictions on religious practice outside officially recognized organizations. Only two Christian organizations—a Catholic church without official ties to Rome and the "Three-Self-Patriotic" Protestant church—are sanctioned by the Chinese government. Unauthorized churches have sprung up in many parts of the country, and unofficial religious practice is flourishing. In some regions, authorities have tried to control activities of these unregistered churches. In other regions, registered and unregistered groups are treated similarly by authorities, and congregations worship in both types of churches. Most Chinese Catholic bishops are recognized by the Pope, and official priests have Vatican approval to administer all the sacraments.

With a population officially just over 1.3 billion and an estimated growth rate of about 0.6, percent, China is very concerned about its population growth and has attempted, with mixed results, to implement a strict birth limitation policy. China's 2002 Population and Family Planning Law and policy permit one child per family, with allowance for a second child under certain circumstances, especially in rural areas, and with guidelines looser for ethnic minorities with small populations. Enforcement varies and relies largely on "social compensation fees" to discourage extra births. Official government policy opposes forced abortion or sterilization, but in some localities, there are instances of forced abortion. The government's goal is to stabilize the population in the first half of the 21st century, and current projections are that the population will peak at around 1.6 billion by 2050.

The next 5 years represent a critical period in China's development. To investors and firms, especially following China's accession to the World Trade Organization (WTO) in 2001, China represents a vast market that has yet to be fully tapped and a low-cost base for export-oriented production. Educationally, China is forging ahead as partnerships and exchanges with foreign universities have helped create new research opportunities for its students. China will host the Summer Olympics in 2008 and views this as an opportunity to showcase to the world its development gains of the past two decades. The new leadership is committed to generating greater economic development in the interior and providing more services to those who do not live in China's coastal areas, goals that form the core of President Hu's concepts of a "harmonious society" and a "spiritual civilization." However, there is still much that needs to change in China. Human rights issues remain a major concern, as does China's lack of effective controls to prevent proliferation of weapons of mass destruction(WMD)-related materials and technology.

The 70.8 million member Chinese Communist Party, authoritarian in structure and ideology, continues to dominate government. Nevertheless, China's population, geographical vastness, and social diversity frustrate attempts to rule by fiat from Beijing. Central leaders must increasingly build consensus for new policies among party members, local and regional leaders, influential non-party members, and the population at large.

In periods of greater openness, the influence of people and organizations outside the formal party structure has tended to increase, particularly in the economic realm. This phenomenon is most apparent today in the rapidly developing coastal region. Nevertheless, in all important government, economic, and cultural institutions in China, party committees work to see that party and state policy guidance is followed and that non-party members do not create autonomous organizations that could challenge party rule. Party control is tightest in government offices and in urban economic, industrial, and cultural settings; it is considerably looser in the rural areas, where the majority of the people live.

The State Department's annual China human rights and religious freedom reports have noted China's well-documented abuses of human rights in violation of internationally recognized norms, stemming both from the authorities' intolerance of dissent and the inadequacy of legal safeguards for basic freedoms. Reported abuses have included arbitrary and lengthy incommunicado detention, forced confessions, torture, and mistreatment of prisoners, as well as severe restrictions on freedom of speech, the press, assembly, association, religion, privacy, worker rights, and coercive birth limitation. In 2005, China stepped up monitoring, harassment, intimidation, and arrest of journalists, Internet writers, defense lawyers, religious activists, and political dissidents. The activities of NGOs, especially those relating to the rule of law and expansion of judicial review, have been curtailed. The Chinese government recognizes five official religions—Buddhism, Islam, Taoism, Catholicism, and Protestantism—and seeks to regulate religious groups and worship. Unregistered religious groups and spiritual movements, as well as religious believers who seek to practice their faith outside of state-controlled religious venues are subject to intimidation, harassment, and detention. In 2004, the Secretary of State again designated China as a "Country of Particular Concern" under the International Religious Freedom Act for particularly severe violations of religious freedom.

At the same time, China's economic growth and reform since 1978 has dramatically improved the lives of hundreds of millions of Chinese, has increased social mobility, and has expanded the scope of personal freedom. This has meant substantially greater freedom of travel, employment opportunity, educational and cultural pursuits, job and housing choices, and access to information. In recent years, China has also passed new criminal and civil laws that provide additional safeguards to citizens. Village elections have been carried out in over 90 percent of China's one million villages.

Since 1979, China has reformed and opened its economy. The Chinese leadership has adopted a more pragmatic perspective on many political and socioeconomic problems and has reduced the role of ideology in economic policy. China's ongoing economic transformation has had a profound impact not only on China but on the world. The market-oriented reforms China has implemented over the past two decades have unleashed individual initiatives and entrepreneurship. The result has been the largest reduction of poverty and one of the fastest increases in income levels ever seen. Today, China is the fourth-largest economy in the world. It has sustained average economic growth of over 9.5 percent for the past 26 years. In 2005, its $2.26 trillion economy was about 1/7 the size of the U.S. economy.

China is the world's most populous country and one of the largest producers and consumers of agricultural products. Roughly half of China's labor force is engaged in agriculture, even though only 10 percent of the land is suitable for cultivation, and agriculture contributes only 13 percent to China's GDP. China's cropland area is only 75 percent of the U.S. total, but China still produces about 30 percent more crops and livestock than the U.S. because of intensive cultivation. China is among the world's largest producers of rice, corn, wheat, soybeans, vegetables, tea, and pork. Major non-food crops include cotton, other fibers, and oilseeds. China hopes to further increase agricultural production through improved plant stocks, fertilizers, and technology. Incomes for Chinese farmers are stagnating, leading to an increasing wealth gap between the cities and countryside. Government policies that continue to emphasize grain self-sufficiency and the fact that farmers do not own—and cannot buy or sell—the land they work have contributed to this situation. In addition, inadequate port facilities and lack of warehousing and cold storage facilities impede both domestic and international agricultural trade.

Industry and construction account for about 46 percent of China's GDP. Major industries are mining and ore processing; iron; steel; aluminum; coal, machinery; textiles and apparel; armaments; petroleum; cement; chemicals; fertilizers; consumer products, including footwear, toys, and electronics; automobiles and other transportation equipment, including rail cars and locomotives, ships, and aircraft; and telecommunications.

China has become a preferred destination for the relocation of global manufacturing facilities. Its strength as an export platform has contributed to incomes and employment in China. The state-owned sector still accounts for about 40 percent of GDP. In recent years, authorities have been giving greater attention to the management of state assets—both in the financial market, as well as among state-owned enterprises—and progress has been noteworthy.

In 2003, China surpassed Japan to become the second-largest consumer of primary energy after the United States. China is also the third-largest energy producer in the world, after the United States and Russia. China's electricity consumption is expected to grow by over 4 percent a year through 2030, which will require more than $2 trillion in electricity infrastructure investment to meet the demand. China expects to add approximately 15,000 megawatts of generating capacity a year, with 20 percent of that coming from foreign suppliers.

Coal makes up the bulk of China's energy consumption (64 percent in 2002), and China is the largest producer and consumer of coal in the world. As China's economy continues to grow, China's coal demand is projected to rise significantly. Although coal's share of China's overall energy consumption will decrease, coal consumption will continue to rise in absolute terms.

Announced in 2005, the 11th Five-Year Program calls for greater energy conservation measures, including the development of renewable energy sources and increased attention to environmental protection. Moving away from coal toward cleaner energy sources, including oil, natural gas, renewable energy, and nuclear power, is an important component of China's development program. China has abundant hydroelectric resources; the Three Gorges Dam, for example, will have a total capacity of 18 gigawatts when fully on-line (projected for 2009). Additionally, the share of electricity generated by nuclear power is projected to grow from 1 percent in 2000 to 5 percent in 2030. China's renewable energy law, which went into effect in 2006, calls for 10 percent of its energy to come from renewable energy sources by 2020.

Since 1993, China has been a net importer of oil, a large portion of which comes from the Middle East. Net imports are expected to rise to 3.5 million barrels per day by 2010. China is interested in diversifying the sources of its oil imports and has invested in oil fields around the world. Beijing also plans to increase China's natural gas production, which currently accounts for only 3 percent of China's total energy consumption. Analysts expect China's consumption of natural gas to more than double by 2010.

One of the serious negative consequences of China's rapid industrial development has been increased pollution and degradation of natural resources. A World Health Organization report on air quality in 272 cities worldwide concluded that seven of the world's 10 most polluted cities were in China. According to China's own evaluation, two-thirds of the 338 cities for which air-quality data are available are considered polluted—two-thirds of them moderately or severely so. Respiratory and heart diseases related to air pollution are the leading cause of death in China. Almost all of the nation's rivers are considered to be polluted to some degree, and half of the population lacks access to clean water. By some estimates, approximately 300 million residents drink contaminated water every day. Ninety percent of urban bodies of water are severely polluted. Water scarcity is also an issue; for example, severe water scarcity in Northern China is a serious threat to sustained economic growth, and the government has begun working on a project for a large-scale diversion of water from the Yangtze River to northern cities, including Beijing and Tianjin. Acid rain falls on 30 percent of the country. Various studies estimate pollution costs the Chinese economy seven to ten percent of GDP each year.

China's leaders are increasingly paying attention to the country's severe environmental problems. In 1998, the State Environmental Protection Administration (SEPA) was officially upgraded to a ministry-level agency, reflecting the growing importance the Chinese Government places on environmental protection. In recent years, China has strengthened its environmental legislation and made some progress in stemming environmental deterioration. In 2005, China joined the Asia Pacific Partnership on Clean Development, which brings industries and governments together to implement strategies that reduce pollution and address climate change. During the 10th Five-Year Plan, China plans to reduce total emissions by 10 percent. Beijing, in particular, is investing heavily in pollution control as part of its campaign to host a successful Olympiad in 2008. In recent years, some cities have seen improvement in air quality.

China is an active participant in climate change talks and other multilateral environmental negotiations, taking environmental challenges seriously but pushing for the developed world to help, to a greater extent, developing countries. It is a signatory to the Montreal Protocol for the Protection of the Ozone Layer, to the Convention on International Trade in Endangered Species, to the Basel Convention, which governs the transport and disposal of hazardous waste, and to other major environmental agreements.

The question of environmental impacts associated with the Three Gorges Dam project has generated controversy among environmentalists inside and outside China. Critics claim that erosion and silting of the Yangtze River threaten several endangered species, while Chinese officials say the dam will help prevent devastating floods and generate clean hydroelectric power that will enable the region to lower its dependence on coal, thus lessening air pollution.

Science and technology have always preoccupied China's leaders; indeed, China's political leadership comes almost exclusively from technical backgrounds and has a high regard for science. Deng called it "the first productive force." According to some Chinese science policy experts, distortions in the economy and society that have been created by party rule have severely hurt Chinese science. The Chinese Academy of Sciences, modeled on the Soviet system, puts much of China's greatest scientific talent in a large, under-funded apparatus that remains largely isolated from industry. The reforms of the past decade have begun to address this problem, though.

Chinese science strategists see China's greatest opportunities in newly emerging fields, such as biotechnology and computers, where there is still a chance for China to become a significant player. Most Chinese students that studied abroad have not returned, but they have built a dense network of trans-Pacific contacts that will greatly facilitate U.S.-China scientific cooperation in the coming years. The U.S. space program is often held as the standard of scientific modernity in China. China's small but growing space program, which successfully completed its second manned orbit in October 2005, is a source of national pride.

China's merchandise exports totaled $762.3 billion and imports totaled $660.2 billion in 2004. Its global trade surplus surged from $32 billion in 2004 to $102 billion in 2005. China's primary trading partners include Japan, the EU, the United States, South Korea, Hong Kong, and Taiwan. According to U.S. statistics, China had with the U.S. a trade surplus of $201.6 billion in 2005.

Answers

Concept Applications
1. Claims makers
2. White-collar crimes
3. Secret deviants
4. Falsely accused
5. Mores

Multiple-Choice				**True/False**	
1. c page 168	11. c page 179			1. F page 168	
2. b page 170	12. c page 179			2. T page 160	
3. a page 171	13. c page 179			3. T page 170	
4. a page 172	14. b page 180			4. T page 170	
5. d page 173	15. a page 182			5. F page 173	
6. c page 176	16. d page 185			6. T page 174	
7. b page 176	17. d page 185			7. T page 177	
8. a page 177	18. c page 190			8. F page 179	
9. c page 178	19. a page 191			9. T page 184	
10. d page 179	20. c page 191			10. T page 185	

Chapter 8

Social Stratification

Study Questions

1. As members of the world's richest country, what questions about wealth distribution are we obligated to ask?

2. Why does the social stratification chapter emphasize the world's richest and poorest peoples?

3. What is the connection between social stratification and life chances?

4. Distinguish between achieved and ascribed characteristics.

5. What does *status value* mean?

6. What characteristics distinguish a caste from a class system of stratification?

7. Use infant mortality and consumption patterns as examples of how life chances vary across countries.

8. How do sociologists describe life chances within countries?

9. Explain the basic dynamics of apartheid.

10. Is the United States a class system? Why or why not?

11. Are caste and class systems distinct types of stratification systems? Explain.

12. In what ways is inequality in the United States systematic?

13. How do the "functions of poverty" help us to understand whose needs are being met by a system that pays so many so little for their labor?

14. How do functionalists (Davis and Moore) explain social stratification?

15. Explain the conflict position on Davis and Moore's theory.

16. From a world system perspective, how has capitalism come to dominate the global network of economic relationships?

17. Distinguish among core, peripheral, and semiperipheral economies. Give an example of a country that fits each of these three economies.

18. What are some of the structural responses to global inequality? Are those responses taking place? How effective are those responses?

19. Distinguish between colonialism and neocolonialism.

20. Summarize how Marx approached social class in his writings. What are the contemporary applications for Marx's ideas?

21. How does Max Weber use the concept of social class? What are the contemporary applications?

22. How is class ranking complicated by status groups and parties?

23. What general structural changes in the American economy have created the urban and other underclasses?

Concept Applications

Consider the concepts listed below. Match one or more of the concepts with each scenario. Explain your choices.

    a. Ascribed characteristics
    b. Intergenerational mobility
    c. Life chances
    d. Negatively privileged property class
    e. Social stratification
    f. Status group
    g. Status value
    h. Upward mobility
    i. Vertical mobility

Scenario 1

"Do blondes have more fun? Social scientists have yet to nail down the answer. But economists now have good reason to believe that blondes make more money—or at least the trim, attractive ones do. New studies show that men and women (with any hair color) who are rated by survey interviewers as below average in attractiveness typically earn 10 to 20 percent less than those rated above average.

"One is tempted to write off the results as proof that idle econometricians are the Devil's helpers. But the findings from Daniel Hamermesh of the University of Texas and Jeff Biddle of Michigan State are complemented by other research showing that obese women are also at a considerable earnings disadvantage. And they could figure prominently in the very serious business of deciding who is protected by the three-year-old Americans with Disabilities Act" (Passell 1994:C2).

Scenario 2

"The Brinks Hotel was another American symbol in Saigon. It was a bachelor officer's headquarters, an American world that Vietnamese need not enter unless, of course, it was to clean the rooms or to cook or to provide some other form of service. It stood high over Saigon and its hovels, a world of Americans eating American food, watching American movies, and just to make sure that there was a sense of home, on the roof terrace there was always a great charcoal grill on which to barbecue thick American steaks flown in especially to that end" (Halberstam 1987:618-19).

Scenario 3

"These children [of people who make enough money to live a privileged life] learn to live with choices—more clothes, a wider range of food, a greater number of games and toys—that other boys and girls may never be able to imagine. They learn to grow fond of or resolutely ignore dolls and more dolls, large dollhouses, and all sorts of utensils and furniture to go in [these dollhouses, as well as] enough Lego sets to build yet another house for the adults in the family. They learn to take for granted enormous playrooms filled to the brim with trains, helicopters, boats, punching bags, Monopoly sets.... They learn to assume instruction—not only at school, but at home—for tennis, for swimming, for dancing, for horse riding. And they learn often enough to feel competent at those sports, in control of themselves while playing them, and, not least, able to move smoothly from one to the other" (Coles 1978:26).

112

Scenario 4

"Wanting out is a common ambition in small towns all over America. In 1951, there were three ways to realize it. One was to get a job in the big city—in my case, either Kansas City or St. Louis, at the edges of the imaginable world. At sixteen, I was too young for this, and besides, I had no idea of what I could do.

"A second way—chosen by four men from the class ahead of me—was to enlist in a branch of military service or volunteer for the draft. That would get you even farther from home and pile up educational benefits under the GI Bill.

"A third alternative to work and military service was just beginning to open up to people—mostly men—of my class and region: college" (Davis 1996:14).

Scenario 5

"The deeper message of Edin's book concerns the material hardships that most welfare families still endure. Eight in 10 had severe housing problems. One in six had recently been homeless. One-third had run out of food sometime in the previous year. And conditions didn't really improve for those who appear to have moved up one step to an entry-level job. In examining the budgets of 165 working mothers, Edin found them even more likely than those on welfare to be unable to pay their bills. 'I thought they might be the same, but not worse,' she says" (DePerle 1997:34).

Practice Test: Multiple-Choice Questions

1. Which one of the following countries has the lowest poverty rate in the world?
   a. Mexico
   b. Haiti
   c. Taiwan
   d. North Korea

2. The systematic process by which individuals, groups, and places are ranked on a scale of social worth is
   a. social stratification.
   b. symbolic stratification.
   c. apartheid.
   d. social structure.

3. A baby born in _____ has one of the best chances of surviving its first year of life.
   a. the United States
   b. Sweden
   c. Italy
   d. Singapore

4. Which of the following is an achieved characteristic?
   a. wrinkles
   b. skin color
   c. occupation
   d. reproductive capacity

5. People assign _____ when they regard some features of a characteristic as more valuable or worthy than other features.
   a. life chances
   b. status value
   c. social class
   d. social stratification

6. Caste systems of stratification are characterized by <u>all</u> but which one of the following adjectives?
   a. rigid
   b. closed
   c. restricted
   d. fluid

7. Apartheid is an example of
   a. a caste system of stratification.
   b. a class system of stratification.
   c. an ideal type of social stratification.
   d. intergenerational mobility.

8. A person who changes his or her class position through graduation, inheritance, or job promotions is experiencing
   a. vertical mobility.
   b. horizontal mobility.
   c. caste mobility.
   d. downward mobility.

9. In the United States, concrete finishers and cement masons are likely to be which one the following?
   a. male and white
   b. Hispanic and male
   c. Hispanic and female
   d. white and female

10. In analyzing social stratification, functionalists ask
    a. who benefits from social stratification and at whose expense.
    b. how do people of different social statuses interact.
    c. why some positions in society are more valued than other positions.
    d. why the disadvantaged lack the work ethic needed to advance.

11. The question "Why should full-time workers at a child care center (a traditionally female occupation) receive a median weekly salary of $333, while a person working as an auto mechanic (a traditionally male occupation) earns $578?" relates to issues of
    a. pay equity.
    b. comparable worth.
    c. functional uniqueness.
    d. status consciousness.

12. Many businesses, governmental agencies, and nonprofit organizations exist to serve poor people or to monitor their behavior. This arrangement is an example of
    a. functional uniqueness.
    b. comparable worth.
    c. the functions of poverty.
    d. status consciousness.

13. According to world-system theorists, capitalism has come to dominate the world economy because
    a. under this system, governments control economic activities.
    b. it is the only economic system in the world.
    c. of the ways in which capitalists respond to changes in the economy, especially to economic stagnation.
    d. national interests take precedence over corporate interests.

14. Japan, Germany, the United States, Canada, Italy, and the United Kingdom possess _____ economies.
    a. semiperipheral
    b. core
    c. peripheral
    d. transitional

15. Many successful people view government and society as irrelevant or, worse, a hindrance to their good fortune; instead, they attribute their success to their own character and individual performance. This viewpoint represents the _____ theory of wealth creation.
    a. "great man"
    b. societal-created
    c. narcissistic
    d. free ride

16. _____ is the term for continuing economic dependence on former colonial powers.
    a. Neocolonialism
    b. Social stratification
    c. Conflict
    d. Colonialism

17. The bulk of U.S. foreign assistance to the world's poorest countries goes toward all but which one of the following?
    a. development
    b. crisis intervention
    c. military training and financing
    d. narcotics control

18. _____ is the flow of the most educated people from poor to rich economies.
    a. Subsidized education
    b. Out-migration
    c. In-migration
    d. Brain drain

19. Karl Marx believed that _____ was the most important engine of change.
    a. technology
    b. societal need
    c. class struggle
    d. ideology

20. Weber's ideas about social class inspire sociologists to
    a. study the people who comprise the middle class.
    b. compare the situation of the wealthiest with that of the poorest.
    c. study class conflict as an agent of change.
    d. think of social class as determined by one's relationship to the means of production.

True/False Questions

1. T F Ascribed characteristics are attributes people cannot easily change.

2. T F Inequality exists both across countries and within countries.

3. T F A baby born in the United States has the best chance in the world of surviving its first year of life.

4. T F Apartheid, South Africa's system of social stratification, was abolished (in a legal sense) in 1994.

5. T F In a true class system, there is no inequality.

6. T F In the United States, chances are good that a dental assistant is a female classified as white.

7. T F The Wal-Mart CEO earns 6,000 times the salary of the Chinese factory worker who makes products sold in the CEO's stores.

8. T F The United States is the largest donor in absolute dollars of foreign aid.

9. T F Social class is difficult to define.

10. T F In numerical terms, people classified as black represent the largest racial category living in poverty.

Internet Resources

- **The Bible on the Poor**
  http://www.zompist.com/meetthepoor.html
  "The Bible contains more than 300 verses on the poor, social justice, and God's deep concern for both. This page contains a wide sample of them, and some reflections. It's aimed at anyone who takes the Bible seriously."

- **Debt Relief Under the Heavily Indebted Poor Countries (HIPC) Initiative**
  http://www.imf.org/external/np/exr/facts/hipc.htm
  The HIPC Initiative was intended to resolve the debt problems of the most heavily-indebted poor countries (originally 41 countries, mostly in Africa) with total debt nearing $200 billion. The 600 million people living in these countries survive an average of 7 years less than citizens in other developing countries, with half living on less than $1 per day. Money freed up by debt relief must be used for sustainable development, so that the countries will not again face unmanageable debts and their people can exit from extreme poverty.

- **Rich World, Poor World**
  http://www.cgdev.org/Research/?TopicID=39
  "The Center for Global Development is dedicated to reducing global poverty and inequality through policy-oriented research and active engagement on development issues with the policy community and the public. A principal focus of the Center's work is the policies of the United States and other industrial countries that affect development prospects in poor countries." This site gives more information on CBD research and publications on issues such as *Education and the Developing World, Global Trade, Jobs and Labor Standards* and *Global HIV/AIDS and the Developing World.*

Applied Research

Sociologists Ross E. Mouer and Yoshio Sugimoto (1990) present a multidimensional framework for thinking about stratification. They identify four dimensions of stratification: economic, political, psychological, and information-based, and give examples of rewards associated with each dimension. Select one specific type of reward from the list below and find data from the U.S. Bureau of the Census or other data sources that illustrate patterns of inequality in the United States.

Economic rewards:
> Occupation
> Salary
> Pension
> Benefits
> Environment (quality of surroundings)
> Employment security
> Job safety
> Quality of recreation facilities
> Leisure

Political rewards:
> Influence
> Authority
> Contacts
> Access to guns and tanks
> Control over army or police force
> Votes
> Publicity
> Information and intelligence

Psychological rewards:
Status
Prestige
Honor
Esteem
Fame
Publicity
Recognition
Friends
Conspicuous consumption

Information-based rewards:
Knowledge
Specific skills
Social awareness
Technical know-how
Access to information

Answers

Concept Applications
1. Status value, Life changes, Ascribed characteristics
2. Status group
3. Life changes, Social stratification
4. Upward mobility, Vertical mobility, Intergenerational mobility
5. Negatively privileged property class

Multiple Choice
1. c page 199
2. a page 200
3. b page 202
4. c page 200
5. b page 201
6. d page 205
7. a page 206
8. a page 206
9. b page 207
10. c page 210

11. b page 212
12. c page 209
13. c page 214
14. b page 214
15. a page 213
16. a page 217
17. a page 219
18. d page 215
19. c page 221
20. b page 224

True/False
1. T page 200
2. T page 203
3. F page 202
4. T page 206
5. F page 206
6. T page 207
7. T page 213
8. T page 217
9. T page 220
10. F page 217

C hapter 9

Race and Ethnicity

Study Questions

1. Explain how the peopling of the United States is a global story.

2. Why is it important to focus on the U.S. system of racial and ethnic classification?

3. Is race a biological fact? Explain.

4. What is ethnicity?

5. Name the 6 official single race categories as designated by the U.S. Census Bureau. How did the system of racial classification change for the 2000 Census?

6. What arc the problems associated with assigning people to racial categories?

7. What is the total number of multiple-race categories in the United States? What percentage of Americans identify with more than one race?

8. Describe the U.S. system of ethnic classification. Why is the label *Hispanic* confusing?

9. Is Hispanic a race? Explain.

10. Define chance, choice, and context. How is race a product of these factors?

11. How is ethnicity a product of chance, choice, and context?

12. Why does the United States not include Arab or Middle Eastern as a racial category?

13. Why are people of Middle Eastern and Arab ancestry classified as white?

14. Why is the 2003 Census population brief on Arab populations within the United States historic?

15. What percentage of the U.S. population is foreign-born? Who are the foreign-born?

16. Give two examples of how race and ethnicity have been connected to U.S. immigration policy?

17. Immigration has always inspired debate in the United States. Why?

18. What are minority groups? What are the essential characteristics of all minority groups?

19. Distinguish between absorption assimilation and melting pot assimilation.

20. What are racist ideologies? Give at least two examples showing how racist ideologies are used to justify one group's domination over another.

21. How do systems of racial classification coincide with discrimination?

22. What are the reasons black athletes dominate some sports, such as basketball and are virtually invisible in others?

23. What is a stereotype? How are stereotypes perpetuated and reinforced?

24. According to Robert K. Merton, what is the relationship between prejudice and discrimination?

25. Distinguish between individual discrimination and institutional discrimination. Give examples.

26. What is a stigma? How is this concept relevant to issues of race and ethnicity?

27. What are mixed contacts? How do stigmas dominate the course of interaction between the stigmatized and the "normals"?

28. How do the stigmatized respond to people who treat them as members of a category?

Concept Applications

Consider the concepts listed below. Match one of more of the concepts with each scenario. Explain your choices.

a. Institutionalized discrimination
b. Involuntary minorities
c. Melting pot assimilation
d. Mixed contacts
e. Selective perception
f. Stereotypes
g. Stigma

Scenario 1

"In the book, an American Asian woman finds her white beau attractive because he is from Connecticut, not Canton. He is tall and lanky; he does not have skinny arms like her brothers and father. He is commanding and gets what he wants. Asian men, however, are not depicted as commanding but as arrogant and chauvinistic. My Asian father has never treated my mother arrogantly. He is not short or uncommunicative either. My father is tall with broad shoulders, a physical attribute inherited by both my brother and me. We have his strong jaw line, too. And I have a dimple on my chin like actor Kirk Douglas. An American Asian woman acquaintance made a comment that my brother and I were unlike 'typical' Asian men because we are tall and muscular. Her own brother is tall and muscular! It gets worse. Two strangers from Latin America, on two separate occasions, asked me if I was 'mixed.' Both refused to believe that I was 100% Asian because I did not fit their stereotype of what an Asian should look like. One even referred to my 'big' eyes'" (Wang 1994:20).

Scenario 2

"...[I]ntegration can be seen as a two-way process in which the dominant and subordinate sectors interact to forge a new entity, in much the same way as different paints in a bucket. Under integration, the best elements of both the majority and minority culture are merged into a single and coherent national framework across a range of practices, including intermarriage and education" (Fleras and Elliott 1992:62).

Scenario 3

"Finally, the category 'Native American' is an artifice of the colonial collision. It is composed of multiple socio-cultural groups who share a colonial history as Indians. They, too, were marginalized in and excluded from full and equal participation in mainstream American institutions and practices. First, there was the military conquest of the Native Americans and their subsequent removal to reservations. But, almost from their first interactions, Native Americans sought education from the United States government. In more than one-quarter of the approximately four hundred treaties entered into by the United States government between 1778 and 1871, education was one of the specific services Native Americans requested in exchange for their lands. But in the formalized education provided by the United States, Native American students were forced to embrace Western ideas and culture, whose price was the repression and denial of their own cultures. Many students were forced into a cultural no-man's land where they remained torn between two worlds. Most students simply dropped out of the system" (Hogue 1996:9).

Scenario 4

"The Tuskegee study began in 1932 when the Public Health Service (and later the Centers for Disease Control) decided to follow 400 black men with syphilis without treating them. The subjects, who were recruited from churches and clinics throughout the South, were told only that they had 'bad blood'" (Stryker 1997: E4).

Scenario 5

"As soon as I walked into the students' center, I knew I'd gone to the wrong place. Just about everyone there looked really ethnic-African American, Asian, Native American, Latino. And there I was, this white-looking guy. A few other students looked kind of white, too, but at least their names tags made up for it: last names like 'Chan' or 'Lee' or 'Wong.' What's my last name? Jewish. Great. I stood around feeling really out of place until this other student began talking to me. He was African American. 'So, what are you?' he asked me right away. I was relieved to tell him my mom was Chinese, like I was explaining myself. 'Oh, OK, yeah, you can sort of see it,' he said, after eyeing me carefully. 'But would you look at some of the guys here? I don't know what they're supposed to be. I left a little later and never went back" (Hess 1997).

Practice Test: Multiple-Choice Questions

1. _____ is a vast collectivity of people more or less bound together by shared and selected history, ancestries, and physical features.
   a. Assimilation
   b. Race
   c. Segregation
   d. A minority group

2. In the U.S. today, there are _____ race categories if we include single and multiple race responses.
   a. 63
   b. 103
   c. 213
   d. 13

3. Under the U.S. system of racial classification, people with ancestors from Pakistan and Siberia are expected to identify themselves as
   a. Asian.
   b. Native American.
   c. black.
   d. white.

4. In the United States, "a person of Mexican, Puerto Rican, Cuban, or Central or South American culture or origin" is known as
   a. a conquistador.
   b. a Pacific Islander.
   c. a Hispanic.
   d. non-white.

5. _____ is the larger social setting in which racial and ethnic categories are recognized, constricted, and challenged.
   a. Chance
   b. Context
   c. Choice
   d. Conscious

6. The origins of an ideology that supports the U.S. belief in racial purity can be traced to
   a. slavery.
   b. reconstruction.
   c. the Civil War.
   d. the Jim Crow era.

7. Black Americans, whether they are native-born or immigrants, are pressured to identify as black, not as an ethnic group. This situation reflects the lack of
   a. chance.
   b. choice.
   c. context.
   d. choice in context.

8. The only source of data we have on the size of Arab or Middle Eastern populations in the United States comes from
   a. FBI records.
   b. birth and death certificates.
   c. the ancestry question on the census.
   d. driver's licenses.

9. Which one of the following states is most likely to have the largest percentage of foreign-born living within its borders?
   a. Texas
   b. New Jersey
   c. Florida
   d. California

10. Which characteristic applies to the Arab population?
    a. Arabs represent 30 percent of the population in Dearborn, Michigan.
    b. An estimated 25 percent of the Arab population is Christian.
    c. Approximately 10 million people in the United States have Arab ancestry.
    d. People classified as Iraqi account for over 60 percent of the Arab population in the United States.

128

11. The Bracero Program, which began in 1942, allowed _____ to work legally in the United States to relieve labor shortages in rural areas and to bolster the American work force during World War II.
    a. Mexicans
    b. Italians
    c. Japanese
    d. Africans

12. "When I use checks, credit cards, or cash, I can count on my skin color not to work against the appearance that I am financially reliable." This statement illustrates an example of
    a. privileges that members of dominant groups enjoy and take for granted.
    b. privileges that members of dominant groups have earned.
    c. adaptation to dominant culture.
    d. the workings of capitalist societies.

13. The key characteristic determining minority status is
    a. size relative to the dominant group.
    b. voluntary emigration.
    c. lack of access to and control over valued resources.
    d. physical appearance distinct from the majority of people in the society.

14. Former Los Angeles police chief Daryl Gates argues that many blacks have died from restraining chokeholds because their veins and arteries do not open up as fast as they do on normal people. Sociologists classify this argument as an example of
    a. institutionalized discrimination.
    b. a hate crime.
    c. racist ideology.
    d. discrimination.

15. "All marriages of white persons with Negroes, Mulattos, Mongolians, or Malaya hereafter contracted in the State of Wyoming are and shall be illegal and void." This Wyoming law is an example of _____ laws.
    a. Civil Rights
    b. Jim Crow
    c. Civil War era
    d. Integration

16. Bill notes that black athletes dominate the sport of basketball and uses that as evidence of natural leaping ability. At the same time, he does <u>not</u> use the same kind of logic to explain why white athletes dominate gymnastics. Bill is guilty of
    a. selective perception.
    b. assimilation.
    c. institutionalized discrimination.
    d. non-prejudiced discrimination.

17. _____ is the established and customary way of doing things in society that keeps minority members in a disadvantaged position.
    a. Systematic discrimination
    b. Corporate discrimination
    c. Normative discrimination
    d. Institutionalized discrimination

18. A stigma is considered discrediting because
    a. it damages the possessor's reputation.
    b. it means the person cannot get financial credit.
    c. it overshadows all other attributes that a person might possess.
    d. it draws attention to other positive and negative characteristics.

19. _____ is a situation in which a dominant group defines some subgroup of people in racial or ethnic terms, thereby forcing that subgroup to become, appear, or feel more ethnic than they might otherwise be.
    a. Designated ethnicity
    b. Ethnicity
    c. Foreign ethnicity
    d. Involuntary ethnicity

20. Which one of the following groups is <u>most</u> likely to be a voluntary minority?
    a. Native Americans
    b. Mexican-Americans
    c. Native Hawaiians
    d. Korean-Americans

True/False Questions

1. T F   Under the U.S. system of racial classification, parents and their biological children can belong to different races.

2. T F   Physical boundaries separating one racial category from another are clear and definite.

3. T F Hispanics can be of any race.

4. T F Individual choice regarding race is constrained by chance and context.

5. T F In the United States, the biological facts of ancestry have no bearing on the system of racial classifications.

6. T F Japan and Germany welcome immigrants as permanent residents and citizens.

7. T F Someone born on American soil to two parents with illegal immigration status is considered a U.S. citizen.

8. T F People who belong to a minority group are treated as members of a category.

9. T F Racial classification was the cornerstone of the Jim Crow laws enacted in the 1880s.

10. T F No hate crimes seem to involve blacks attacking whites.

Internet Resources

- **Interracial Voice**
  http://www.webcom.com/~intvoice
  Interracial Voice is an electronic publication issued every other month. It "advocates universal recognition of mixed-race individuals as constituting a separate 'racial' entity and supports the initiative to establish a multiracial category on the 2000 census."

Applied Research

Choose one of the followings books: *The Color of Water: A Black Man's Tribute to His White Mother* by James McBride, *Member of the Club* by Lawrence Otis Graham, *Showing Our Colors: Afro-German Women Speak Out* edited by May Opitz, Katharina Oguntoye and Dagmar Schultz, *Black Indians: A Hidden Heritage* by William Loren Katz, *Border Lands* by Gloria Anqaldua or *Life on the Color Line* by Gregory Howard Williams. Write a book review describing how the book you have chosen supports the idea that race classification as practiced in the U.S. makes no sense.

Answers

Concept Applications
1. Selective perception, Stigma, Mixed contacts, Stereotypes
2. Melting pot assimilation
3. Involuntary minorities
4. Institutionalized discrimination
5. Mixed contacts

Multiple Choice

1.	b	page 234	11.	a	page 244
2.	a	page 236	12.	a	page 248
3.	a	page 235	13.	c	page 247
4.	c	page 237	14.	c	page 252
5.	b	page 239	15.	b	page 253
6.	a	page 240	16.	a	page 253
7.	b	page 239	17.	d	page 258
8.	c	page 241	18.	c	page 258
9.	d	page 243	19.	d	page 250
10.	a	page 240	20.	d	page 250

True/False

1.	T	page 232
2.	F	page 236
3.	T	page 235
4.	T	page 239
5.	T	page 239
6.	F	page 242
7.	T	page 242
8.	T	page 247
9.	T	page 253
10.	F	page 257

hapter 10

Gender

Study Questions

1. Why is American Samoa the focus of a chapter on gender?

2. Distinguish between primary and secondary sex characteristics. Is sex (the biological concept of male and female) a clear-cut distinction?

3. Define gender. Why do sociologists find the concept of "gender" useful?

4. What is gender polarization? Give an example.

5. Define gender-schematic and gender polarization. How are they connected? Explain how gender-schematic decisions and sexual desire affect educational choices and sexual desire.

6.   Explain the following statement: "People of the same sex vary in the extent to which they meet their society's gender expectations."

7.   What are *fa-afafines*? What characteristics of Samoan society support this gender blurring?

8.   How does socialization operate to teach people about society's gender expectations?

9.   What socialization mechanisms are at work in American Samoa to encourage interest and success in football among males?

10. What does the commercialization of gender ideals mean? Give at least one example.

11. How do structural constraints help to explain male-female differences? Specifically, how does one's position in the social structure channel behavior in stereotypically male or female directions?

12. What is sexist ideology? How is sexist ideology reflected in military policy toward homosexuals?

13. What is feminism? Give examples of the variety of feminist positions.

14. What is ethgender?

15. Name five areas of women's lives over which the state may choose to exercise control.

Concept Applications

Consider the concepts listed below. Match one or more of the concepts with each scenario. Explain your choices.

    a. Femininity (or feminine characteristics)
    b. Gender
    c. Gender polarization
    d. Structural constraints
    e. Intersexuals

Scenario 1

"Women were widely excluded from the jury service until a few decades ago, many years after it was no longer permissible to exclude blacks as a group, and it was not until 1975 that the Supreme Court ruled that states had to maintain a representative jury pool that included women" (Greenhouse 1994:A10).

Scenario 2

High-heeled shoes are still meant predominantly for posing, as Miss America does in her swimsuit. She keeps her legs together, one knee gently bent. Pictures of women in bathing suits with heeled legs astride make a more up-to-date, but not necessarily a more feminist, statement.

High heels have never been made for comfort or for ease of movement. Their first wearers spoke of themselves as "mounted" or "propped" upon them; they were strictly court wear and constituted proof that one intended no physical exertion and need make none.

The Chinese had long known footwear that had the same effect, with wooden pillars under the arch of each shoe so that wearers required one or even two servants to help them totter along. Women had their feet deformed, by binding them into tiny, almost useless fists, which were shod in embroidered bootees. (Visser 1994:38)

Scenario 3

"Women in professional jobs have workplace issues like the glass ceiling and the mommy track. But now there is one for secretaries: rug-ranking. 'If the secretary's pay is based on her boss' status, not on the content of her job, that's rug-ranking—treating her as a perk like the size of his office or the quality of the carpet on his floor,' said N. Elizabeth Fried, a labor consultant based in Dublin, Ohio. 'Secretaries are the only ones in the corporate world whose pay is directly linked to the boss. Instead of a career path of their own, most secretaries have had a hitch-your-wagon-to-a-star reward system" (Lewin 1994:A1).

Scenario 4

"Family work was structured around gender and age…Women were responsible for the farmyard economy of milking, rearing of young animals, poultry, butter making, and frequently the cultivation of vegetables as well…Animal husbandry, the buying and selling of animals, and most fieldwork (e.g., plowing, burrowing) and structural yard work (e.g., building, repairing) requiring heavy effort was undertaken by the male 'farmer'" (O'Hara 2001).

Scenario 5

"[There] are rare cases in which babies are born whose sexual gender is ambiguous or indeterminate...Sexually ambiguous infants, who either appear to be female but are biologically male or appear to be male but are biologically female are sometimes called pseudohermaphrodites" (Scarboro 1991:339).

Practice Test: Multiple-Choice Questions

1. American Samoa is the focus of the gender chapter because that society
   a. makes the greatest distinctions between males and females.
   b. makes the least distinctions between males and females.
   c. channels male energy into pursuing football careers.
   d. channels female energy into pursuing cheerleading careers.

2. A person's sex is determined first and foremost on the basis of
   a. distribution of facial and body hair.
   b. secondary sex characteristics.
   c. primary sex characteristics.
   d. social expectations.

3. From a sociological point of view, sex is _____ and gender is _____.
   a. a biologically based classification scheme; a socially constructed phenomenon
   b. a socially constructed phenomenon; a feminist creation
   c. a classification scheme; a continuum
   d. a socially constructed phenomenon; a biological construction

4. Spanish hurdler Maria Jose Martinez Patino lost her right to compete in amateur and Olympic events because she failed the sex tests. Martinez challenged that decision. In the end, the IAAF ruled that Martinez
   a. was a "man."
   b. was intersexed.
   c. would have to compete with men in the future.
   d. possessed no special advantages over other female competitors.

5. _____ are physical traits not essential to reproduction.
   a. Primary sex characteristics
   b. Secondary sex characteristics
   c. Chromosomal sex characteristics
   d. Genders

6. _____ is the physical, behavioral, and mental or emotional characteristics believed to be characteristic of males.
   a. A secondary sex characteristic
   b. Gender
   c. Masculinity
   d. A primary sex characteristic

7. In pre-Christian Samoa, ideal standards signifying manhood centered around
   a. speed and agility.
   b. hair.
   c. body tattooing.
   d. muscles.

8. Sandra Bem wrote that a person's sex is connected to "virtually every aspect of human experience, including modes of dress, social roles, and even ways of expressing emotion and experiencing sexual desire." Bem was writing about
   a. primary sex characteristics.
   b. gender polarization.
   c. genetic and chromosomal sex.
   d. the intersexed.

9. A woman consciously or unconsciously decides that the man she dates must be bigger, taller, and stronger than she is. Her decision can be classified as
   a. feminist.
   b. natural.
   c. anti-feminist.
   d. gender-schematic.

10. Michael notes, "There is an unwritten rule my friends and I follow: if you are going to touch me, make it hurt." Michael is identifying a
    a. rite of passage.
    b. primary sex characteristic.
    c. feeling rule.
    d. universal norm.

11. In American Samoa, *fà-afafines* are
    a. biological females who have taken on "men's ways."
    b. males with tattoos covering the lower body.
    c. females with tattoos covering the lower body.
    d. males who have taken on the "way of women."

12. The two largest employers in American Samoa are
    a. the military and banks.
    b. tourism and tuna canneries.
    c. the government and tuna canneries.
    d. GM and McDonald's.

13. Which one of the following statements about early childhood teachers is false?
    a. Teachers are more accepting of girls' cross-gender behavior and explorations than of boys' cross-gender behaviors and exploration.
    b. Teachers believe that boys who behave like sissies are at greater risk than girls of growing up to be homosexuals.
    c. Teachers are more accepting of boys' cross-gender behaviors and exploration than of girls' cross-gender behavior and exploration.
    d. Teachers believe that girls who behave like tom-boys are at less risk of growing up to be homosexual.

14. Mattel markets "Barbie" as a(n)
    a. aspirational doll.
    b. doll with traditional values.
    c. doll with feminist values.
    d. intersexed toy.

15. Which one of the following learned body language characteristics applies to males?
    a. sitting or standing with legs positioned away from the body.
    b. affiliative facial expression
    c. sitting with legs crossed at the ankles
    d. lowered gaze and constricted body

16. _____ is/are the established and customary rules, policies, and day-to-day practices that affect a person's life chances.
    a. Structural constraints
    b. Ideologies
    c. Selective perceptions
    d. Ethgender

17. Sociologist Renee R. Anspach found that physicians caring for babies in neonatal intensive care units tended to _____ to determine how well infants were doing.
    a. draw on technical information
    b. look for interactional clues
    c. consider a baby's level of alertness
    d. consider a baby's responsiveness to touch

18. According to a Zogby poll, ___ percent of U.S. soldiers from Iraq and Afghanistan indicated that they were extremely uncomfortable interacting with gay colleagues.
    a. 5
    b. 20
    c. 40
    d. 75

19. Sociologists take a_____ perspective when they emphasize in their research and teaching such themes as a right to bodily integrity and autonomy, access to safe contraceptives, the right to choose, and freedom from sexual harassment.
   a. sexist
   b. structuralist
   c. feminist
   d. socialist

20. If we take a long view of women's wages (a 15-year period), we find that the average woman earns _____ percent of what men do during that same time period.
   a. 38
   b. 50
   c. 65
   d. 82

True/False Questions

1. T F   Biological sex is not a clear-cut category.

2. T F   The biological father's contribution of an x or a y chromosome determines the baby's sex.

3. T F   Ninety percent of Bachelor's degrees in Library Science are awarded to males.

4. T F   As a group, males have a longer life expectancy than females.

5. T F   The *fà-afafines* in American Samoa imitate popular foreign female vocalists, such as Britney Spears or Madonna.

6. T F   In American Samoa, the two largest employers are the tuna canneries and the U.S. government.

7. T F   A lowered gaze and constricted body signal deference.

8. T F   In the United States, there is a commercial product on the market to improve almost every female body part or body function.

9. T F   The fertility rate of women on public assistance is higher than the fertility rate of women not on such assistance.

10. T F   The British military recruits soldiers at gay pride events.

Internet Resources

- **U.S. Department of Labor, Bureau of Labor Statistics**
  http://www.dol.gov
  If you are interested gender-related labor statistics visit this government website which posts occupational categories by number and percentage of males and females. You can also find median income by sex for most occupational categories

- **Samoa News**
  http://www.samoanews.com/
  This website posts the most significant stories published in the daily edition of *Samoa News* as well as headline news from past issues.

Applied Research

Use the Google News or some other NewsTracker service to monitor articles generated from the search terms "gender" for one to two weeks. Write a paper describing the major themes of these articles.

Background Notes: Samoa

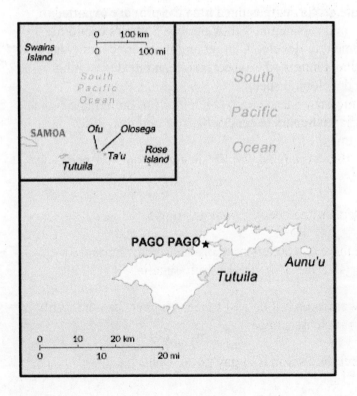

Size: Slightly larger than Washington, DC

Capital: Pago Pago
Population (2006): 57,663
Ethnic groups: Indian-Spanish (mestizo) 60%, Indian 30%, Caucasian 9%, other 1%.
Religions: Christian Congregationalist 50%, Roman Catholic 20%, Protestant and other 30%
Infant mortality rate: 8.88/1000.
Life expectancy: male 73.0 years; female 80 years.
Work force (2005): 17,630
Workforce by Sector: Agriculture--34.0%; services--33%; manufacturing—33.0%;

Settled as early as 1000 B.C., Samoa was "discovered" by European explorers in the 18th century. International rivalries in the latter half of the 19th century were settled by an 1899 treaty in which Germany and the US divided the Samoan archipelago. The US formally occupied its portion - a smaller group of eastern islands with the excellent harbor of Pago Pago - the following year.

Pago Pago has one of the best natural deepwater harbors in the South Pacific Ocean, sheltered by shape from rough seas and protected by peripheral mountains from high winds; strategic location in the South Pacific Ocean.

American Samoa has a traditional Polynesian economy in which more than 90% of the land is communally owned. Economic activity is strongly linked to the US with which American Samoa conducts most of its commerce. Tuna fishing and tuna processing plants are the backbone of the private sector, with canned tuna the primary export. Transfers from the US Government add substantially to American Samoa's economic well being. Attempts by the government to develop a larger and broader economy are restrained by Samoa's remote location, its limited transportation, and its devastating hurricanes. Tourism is a promising developing sector.

Source: Excepted from U.S. Central Intelligence Agency, *World Factbook*, 2007
www.state.gov/r/pa/ei/bgn/35749.htm

Answers

Concept Applications
1. Structural constraints
2. Femininity (or feminine characteristics)
3. Structural constraints
4. Gender polarization
5. Intersexuals

Multiple Choice		**True/False**	
1. c page 267	11. d page 276	1. T page 268
2. c page 268	12. c page 277	2. T page 268
3. a page 269	13. c page 278	3. F page 274
4. d page 269	14. a page 279	4. F page 273
5. b page 269	15. a page 279	5. T page 276
6. c page 270	16. a page 283	6. T page 277
7. c page 271	17. a page 284	7. T page 281
8. b page 272	18. a page 28	8. T page 282
9. d page 273	19. c page 289	9. F page 293
10. c page 274	20. a page 289	10. T page 287

Chapter 11

Economics and Politics

Study Questions

1. Why focus on Iraq in conjunction with the topics of economics and politics?

2. Define economic system. Name three revolutions that have shaped economic systems.

3. Why are the domestication of plants and animals and the invention of the scratch plow considered revolutionary?

4. Name one of the most fundamental features of the Industrial Revolution. Why is this feature fundamental?

5. What do we mean when we say that the Industrial Revolution cannot be separated from European colonization? Use oil as an example.

6. What is a Postindustrial society?

7. What characteristics distinguish a capitalist economic system from a socialist one?

8. From a world system perspective, how has capitalism come to dominate the global network of economic relationships?

9. Distinguish among core, peripheral, and semiperipheral economies.

10. Explain why Iraq would be considered a peripheral economy.

11. Which country is considered to have the strongest and most diverse economy in the world? Why?

12. What contributions do the primary, secondary, and tertiary sectors of the U.S. economy make to the GDP? Which sector contributes the largest share? Explain.

13. What is the difference between a monopoly and an oligopoly?

14. What is a conglomerate? Give an example.

15. What do we mean when we say that the United States is an oil- and mineral-dependent economy?

16. Describe the various kinds of debt that exist in the U.S. economy.

17. Define political system.

18. What is authority? How many types of authority did Weber identify? Give examples of each kind of authority.

19. What are the essential characteristics of a democracy?

20. How do we distinguish between totalitarian and authoritarian governments?

21. What is a theocracy?

22. What is the power elite? Who comprises the power elite in the United States?

23. Does C. Wright Mills believe that there are any significant constraints on the decision-making powers of the power-elite? Why or why not?

24. Explain the pluralist model of power.

25. What are PACs and 527 groups? Give examples.

26. Define empire, imperialistic power, hegemony, and militaristic power.

27. What are some examples of U.S. power and influence in the world?

28. What are insurgents?

Concept Applications

Consider the concepts listed below. Match one or more of the concepts with each scenario. Explain your choices.

   a. Semiperipheral economy
   b. Peripheral economy
   c. Conglomerate
   d. Primary sector
   e. Secondary sector
   f. Special interest groups

Scenario 1

   "In less than three decades, Taiwan has become a major economic player not only in the economy of the Pacific Rim but [also] in the global system as well. Foreign investors have played a vital role in Taiwan's economic development. For example, a mass buyer, like Sears or K-Mart, would visit Taiwanese factories and order goods in bulk for sale under the chain's brand name. A company like Arrow shirts or U.S. Shoe would supply samples to several factories and then contract with the factory that offered the best deal in terms of cost and quality. The "Made in Taiwan" label spread worldwide, even if no one outside Taiwan knew a single Taiwanese company that produced the products" (Goldstein 1991).

Scenario 2

At the urging of Chiquita Brands, a unit of the American Financial Corporation and the world's largest banana producer, the Clinton Administration is seeking to overturn an agreement that guarantees small Caribbean banana farmers special access to the European Union market.

"Why is America doing this to us?" Mr. Prosper, 53, asked as his crop was being boxed at a weighing station here the other day. "This is a little place, and this is all that we know and what we depend on. We have nothing else and we hurt nobody, but now they want to take even this from us."

Much as in neighboring Dominica and St. Vincent and the Grenadines, one-quarter of the labor force in this country of 145,000 people is employed in the banana industry, either growing, processing, or shipping the fruit. In contrast to Central America, where workers paid as little as $2 a day grow most of Chiquita's bananas, Caribbean banana workers are mostly independent growers who own the small plot they farm (Rohter 1997:A6).

Scenario 3

A notable feature of the biggest recent mergers is that the firms dominating this process are not media companies in a strict sense: Disney is avowedly a "family entertainment communication company" in the business of selling theme parks, toys, movies, and videos. Its focus, in the words of CEO Michael Eisner, is on the provision of "non-political entertainment and sports." Time Warner has a similarly wide spectrum of business interests and a comparable marketing orientation. Disney and Time Warner are what Herbert Schiller calls "pop cultural corporate behemoths."

Westinghouse, by contrast, has long been primarily a nuclear power and weapons producer, with its media interests generating only 10 percent of sales revenue. With the Westinghouse takeover of CBS, two of the three top networks are controlled by large firms in the politically sensitive nuclear power/weapons industries (the other is NBC's owner General Electric, which along with Westinghouse is one of the top 15 U.S. defense contractors). (Herman 1996)

Scenario 4

"A mainstay of the mining industry is gold, which is being extracted from the West faster than ever before, says France. About 85 percent of the gold extracted in the West ends up in jewelry, the rest going into products such as electronics." (DiSilvesto 1996)

Scenario 5

"As more private sector organizations learn to use the tools of the political campaign industry, a broad range of corporations, associations, unions, and non-profits are playing a larger, more aggressive role in the shaping of public opinion on matters they deem important.... A study conducted by the Annenberg Public Policy Center of the University of Pennsylvania estimated that during the 1995-96 election cycle, one-third of the total dollars spent on advertising in federal elections was attributable to "issue" advocacy efforts." (Faucheux 1998)

Practice Test: Multiple-Choice Questions

1. The Bush administration invaded Iraq with the intention of establishing a _____ in the heart of the Middle East.
    a. democracy
    b. a totalitarian state
    c. a centrally planned economy
    d. communist government

2. Which one of the following would be classified as "goods"?
    a. entertainment
    b. transportation
    c. financial services
    d. clothing

3. The innovation that turned the hand loom into a power loom, the horse-drawn carriage into the steam engine, and the blacksmith's hammer into a power machine is
    a. capitalism.
    b. domestication.
    c. mechanization.
    d. the computer chip.

4. Under a system of private ownership, _____ own the means of production.
   a. individuals
   b. unions
   c. governments
   d. communes

5. Iraq's economy is heavily dependent on
   a. agriculture.
   b. the service sector.
   c. oil revenues.
   d. migrant labor.

6. Which one of the following is a characteristic of an extremely oil revenue dependent economy?
   a. strong domestic manufacturing sector
   b. a currency that is undervalued
   c. heavily taxed citizens
   d. corruption and political rivalries among those controlling the oil

7. The primary sector of the economy includes economic activities
   a. that generate or extract raw materials from the natural environment.
   b. that transform raw materials into manufactured goods.
   c. related to delivering services.
   d. related to the creation and distribution of information.

8. Which sector of the economy contributes the most to the GDP of the United States?
   a. primary
   b. secondary
   c. tertiary
   d. manufacturing

9. Chris works as a customer service representative. Her job is in the _____ sector of the economy.
   a. primary
   b. secondary
   c. tertiary
   d. peripheral

10. In the United States, union membership varies by state. Which one of the following states has the greatest percentage of workers represented by unions?
    a. Hawaii
    b. Florida
    c. Kentucky
    d. Ohio

11. The worldwide demand for oil and minerals is expected to increase because two countries, which are home to 2.5 billion people, are experiencing dramatic economic growth. These two countries are
    a. Brazil and Mexico.
    b. Iraq and Vietnam.
    c. India and China.
    d. Japan and Canada.

12. The two top foreign holders of U.S. debt are
    a. Spain and South Africa.
    b. Germany and Italy.
    c. Mexico and Canada.
    d. China and Japan.

13. A "chief," "king," or "queen" possesses power based on which form of authority?
    a. traditional
    b. charismatic
    c. legal-rational
    d. socialistic

14. _____ is a system of government in which power is vested in the citizen body or "the people."
    a. Capitalism
    b. Democracy
    c. Totalitarianism
    b. Authoritarianism

15. _____ is a form of government in which political authority is in the hands of religious leaders or a theologically trained elite.
    a. Theocracy
    b. Democracy
    c. Totalitarianism
    d. Authoritarianism

16. C. Wright Mills wrote, "The power to make decisions of national and international consequence is now so clearly seated in political, military, and economic institutions that other areas of society seem off to the side." Mills was writing about
    a. monopolies.
    b. the power elite.
    c. a pluralist society.
    d. conglomerates.

17. In the United States, the National Association of Realtors, the National Auto Dealers Association, and the Association of Trial Lawyers of America contribute to political campaigns and are known as
    a. monopolies.
    b. primary sector industries.
    c. special interest groups.
    d. political action committees.

18. The _____ model views politics as an arena of compromises, alliances, and negotiations among many competing special interest groups.
    a. power elite
    b. pluralist
    c. socialist
    d. capitalist

19. _____ are groups that participate in armed rebellion against an established authority, government, or administration with the hope that those in power will pull out.
    a. Colonizers
    b. PACs
    c. Revolutionaries
    d. Insurgents

20. One U.S. Pentagon official pointed out in a news briefing that "the sheer size of this campaign has never been seen before, never been contemplated before." *This* campaign was the
    a. 1990 oil embargo against Iraq.
    b. Food-for-Oil program in Iraq.
    c. shock and awe campaign in Iraq.
    d. democratization of Iraq.

True/False Questions

1. T F When invading Iraq, the Bush administration was clear on its intention to change the political system in Iraq from a dictatorship to a democracy.

2. T F Many agricultural revolutions have taken place over the course of human history.

3. T F No economic system, even the U.S. system, fully realizes capitalist principles.

4. T F Because of its oil wealth, Iraq is classified as a core economy.

5. T F The U.S. has the most powerful, diverse, and technologically advanced economy in the world.

6. T F China has the largest and most technologically powerful economy in the world.

7. T F A monopoly exists when a handful of producers dominate a market.

8. T F In the United States, Hispanics stand out as the category most likely to vote.

9. T F P.A.C. stands for Political Action Committee.

10. T F Israel is the largest buyer of arms in the world.

Internet Resources

- **Global Policy Forum; Iraq**
  http://www.globalpolicy.org/security/issues/irqindx.htm
  Global Policy Forum's mission is "to monitor policy making at the United Nations, promote accountability of global decisions, educate and mobilize for global citizen participation, and advocate on vital issues of international peace and justice." Its section on Iraq contains links to articles related to Iraq's humanitarian crisis, permanent U.S. bases, and withdrawal.

- **Iraq Daily**
  http://www.iraqdaily.com/
  This website reports daily information on events in Iraq from the World News Network.

- **Iraq Maps**
  http://www.lib.utexas.edu/maps/iraq.html
  This site contains country, city, thematic and detailed maps of Iraq produced by the U.S. Central Intelligence Agency.

Applied Research

Visit the U.S. Department of Defense website that each day announces military contracts worth $5 million or more. Select the latest 50 contracts and map the geographic location of the corporation receiving the contract. Can you draw any conclusions about the U.S. economy from doing this exercise?

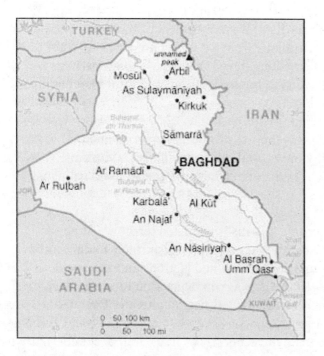

Size: About three times the size of California
Capital: Baghdad City (5.7 million)
Population (2006): 27.5 million.
Ethnic groups: Arab 75%-80%, Kurd 15%-20%, Turcoman, Chaldean, Assyrian, or others less than 5%.
Infant mortality rate: 47.04/1000.
Life expectancy: male 68 years; female 70 years.
Work force (2005): 7.4 million:
Workforce by Sector: Agriculture, forestry, hunting, fishing—21.0%; services— 32.2%; commerce—16.9%; manufacturing—18.7%; construction—5.6%; transportation and communication—4.5%; mining and quarrying—1.0%

GEOGRAPHY
Iraq is bordered by Kuwait, Iran, Turkey, Syria, Jordan, and Saudi Arabia. Almost 75 percent of Iraq's population live in the flat, alluvial plain stretching southeast from Baghdad and Basrah to the Persian Gulf. The Tigris and Euphrates Rivers carry annually about 70 million cubic meters of silt to the delta. Known in ancient times as Mesopotamia, the region is the legendary locale of the Garden of Eden. The ruins of Ur, Babylon, and other ancient cities are in Iraq.

Iraq's two largest ethnic groups are Arabs and Kurds. Other distinct groups are Assyrians, Persians, and Armenians. Arabic is the most commonly spoken language.

155

Kurdish is spoken in the north, and English is the most commonly spoken Western language.

The majority (60-65 percent) of Iraqi Muslims are members of the Shi'a sect, but there is a large (32-37 percent) Sunni population as well, which is comprised of both Arabs and Kurds. Small communities of Christians, Jews, Bahais, Mandaeans, and Yezidis also exist. Most Kurds are Sunni Muslim but differ from their Arab neighbors in language, dress, and customs.

Once known as Mesopotamia, Iraq was the site of flourishing ancient civilizations, including the Sumerian, Babylonian, and Parthian cultures. Muslims conquered Iraq in the seventh century A.D. In the eighth century, the Abassid caliphate established its capital at Baghdad.

At the end of World War I, Iraq became a British-mandated territory. When it was declared independent in 1932, the Hashemite family, which also ruled Jordan, ruled as a constitutional monarchy. In 1945, Iraq joined the United Nations and became a founding member of the Arab League. In 1956, the Baghdad Pact allied Iraq, Turkey, Iran, Pakistan, and the United Kingdom and established its headquarters in Baghdad.

The Iran-Iraq war (1980-88) devastated the economy of Iraq. Iraq declared victory in 1988 but actually achieved a weary return to the status quo antebellum. The war left Iraq with not only the largest military establishment in the Gulf region but also huge debts and an ongoing rebellion by Kurdish elements in the northern mountains. The government suppressed the rebellion by using weapons of mass destruction on civilian targets, including a mass chemical weapons attack on the city of Halabja that killed several thousand civilians.

Iraq invaded Kuwait in August 1990, but a U.S.-led coalition acting under United Nations (UN) resolutions expelled Iraq from Kuwait in February 1991. After the war, Kurds in the north and Shi'a Muslims in the south rebelled against the government of Saddam Hussein. The government responded quickly and with crushing force, killing thousands. It also pursued damaging environmental and agricultural policy meant to drain the marshes of the south. As a result, the United States, the United Kingdom, and France established protective no-fly zones in northern and southern Iraq. In addition, the UN Security Council required the Ba'ath regime to surrender its weapons of mass destruction (WMD) and to submit to UN inspections. When the Ba'ath regime refused to fully cooperate with the UN inspections, the Security Council employed sanctions to prevent further WMD development and to compel Iraqi adherence to international obligations. Coalition forces enforced no-fly zones in southern and northern Iraq to protect Iraqi citizens from attack by the regime and enforced a no-drive zone in southern Iraq to prevent the regime from massing forces to threaten or again invade Kuwait.

A U.S.-led coalition removed the Ba'th regime in March-April 2003, bringing an end to more than 12 years of Iraqi defiance of UN Security Council resolutions.

Today, Iraq is a constitutional democracy with a federal system of government. Since March 2006, the Government of Iraq has been a broad coalition led by a Shi'ite legislative bloc known as the United Iraqi Coalition (UIC) or the United Iraqi Alliance (UIA). The UIC currently holds 128 of 275 seats in the Council of Representatives. The UIC is currently composed of ISCI, the al-Sadr movement, al-Da'wa al-Islamiyya, Da'wa Tanzim al-Iraq, Jama'at al-Fadilah, and various independents. Politicians with Sunni religious affiliations, including the Tawaffuq and Hewar groups, presently hold 59 seats in the Council of Representatives. The Kurdish bloc, known as the Democratic Patriotic Alliance of Kurdistan (which includes the KDP & PUK), holds 53 legislative seats. Ayad Allawi's Iraqiyya or Iraqi National List (INL) holds 25 seats. The remaining seats are composed of various independents.

With regard to the executive branch, much care has been given to ensure that there is proportionate distribution of ministerial positions among the major political groups. For example, in the Presidency Council, President Jalal Talabani is Kurdish, Deputy President 'Adil 'Abd al-Mahdi is a Shi'a Muslim, and Deputy President Tariq al-Hashimi is a Sunni Muslim. Additionally, the Council of Ministers consists of 18 Shi'a Muslims, 8 Sunni Muslims, 8 Kurds, and 5 members of Ayad Allawi's secular INA.

The Government of Iraq is currently working toward reviewing the Constitution. The process is likely to be a long and careful one, as consideration needs to be given to the interests of each of the major political groups. Issues to be addressed include federalism, the sharing of oil revenues, de-Ba'thification reform, and provincial elections.

Historically, Iraq's economy was characterized by a heavy dependence on oil exports and an emphasis on development through central planning. Prior to the outbreak of the war with Iran in September 1980, Iraq's economic prospects were bright. Oil production had reached a level of 3.5 million barrels per day, and oil revenues were $21 billion in 1979 and $27 billion in 1980. At the outbreak of the war, Iraq had amassed an estimated $35 billion in foreign exchange reserves.

The Iran-Iraq war depleted Iraq's foreign exchange reserves, devastated its economy, and left the country saddled with a foreign debt of more than $40 billion. After hostilities ceased, oil exports gradually increased with the construction of new pipelines and the restoration of damaged facilities. Iraq's invasion of Kuwait in August 1990, subsequent international sanctions, damage from military action by an international coalition beginning in January 1991, and neglect of infrastructure drastically reduced economic activity. Government policies of diverting income to key supporters of the regime while sustaining a large military and internal security force further impaired finances, leaving the average Iraqi citizen facing desperate hardships.

Implementation of a UN Oil-For-Food (OFF) program in December 1996 improved conditions for the average Iraqi citizen. In December 1999, Iraq was authorized to export unlimited quantities of oil through OFF to finance essential civilian needs, including food, medicine, and infrastructure repair parts. The drop in GDP in 2001-02 was largely the result of the global economic slowdown and lower oil prices.

Per capita food imports increased significantly, while medical supplies and health care services steadily improved. The occupation of the U.S.-led coalition in March-April 2003 resulted in the shutdown of much of the central economic administrative structure. The rebuilding of oil infrastructure, utilities infrastructure, and other production capacities has proceeded steadily since 2004 despite attacks on key economic facilities and continuing internal security incidents. Despite uncertainty, Iraq is making progress toward establishing the laws and institutions needed to make and implement economic policy.

Iraq's economy is dominated by the oil sector, which has traditionally provided about 95 percent of foreign exchange earnings. Current estimates show that oil production averages 2.0 million bbl/day.

The Iran-Iraq War ended with Iraq sustaining the largest military structure in the Middle East, with more than 70 divisions in its army and an air force of over 700 modern aircraft. Losses during the 1990 invasion of Kuwait and subsequent expulsion of Iraqi forces from Kuwait in 1991 by a UN coalition resulted in the reduction of Iraq's ground forces to 23 divisions and air force to less than 300 aircraft.

When major combat operations ended in April 2003, the Iraqi Army disintegrated, and its installations were destroyed by pilfering and looting. The Coalition Provisional Authority (CPA) officially dissolved the Iraqi military and Ministry of Defense on May 23, 2003. On August 7, 2003, the CPA established the New Iraqi Army as the first step toward the creation of the national self-defense force of post-Saddam Hussein Iraq. Support for the manning, training, and equipping of Iraq's security forces is led by the Multi-National Security Transition Command-Iraq (MNSTC-I). In addition to defense forces, the Ministry of Interior, with the help of the MNSTC-I, is training and equipping civilian police forces to establish security and stability, primarily through combating the nation-wide insurgency. Initially under the command and control of the Multi-National Forces-Iraq (MNF-I) command, police and Iraqi Army units began to transition to Iraqi civilian control in 2006.

Answers

Concept Applications
1. Semiperipheral economy
2. Peripheral economy
3. Conglomerate
4. Primary sector, Secondary sector
5. Special interest groups

Multiple Choice

1.	a	page 299	11.	c	page 312
2.	d	page 300	12.	d	page 313
3.	c	page 301	13.	a	page 314
4.	a	page 305	14.	b	page 315
5.	c	page 307	15.	a	page 318
6.	d	page 307	16.	b	page 320
7.	a	page 309	17.	d	page 322
8.	c	page 310	18.	b	page 321
9.	c	page 310	19.	d	page 324
10.	a	page 310	20.	c	page 324

True/False

1.	T	page 299
2.	T	page 300
3.	T	page 304
4.	F	page 307
5.	T	page 308
6.	F	page 308
7.	F	page 309
8.	F	page 316
9.	T	page 322
10.	F	page 324

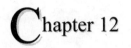hapter 12

Family and Aging

Study Questions

1. Why is Japan the focus of a chapter on family and aging? How does the U.S. compare to Japan on indicators related to family well-being and stability?

2. Why is "family" a difficult concept to define? What are some criteria that might be used to define family?

3. How does the family contribute to order and stability in society? What are some problems with defining family in terms of social functions?

4. What is the conflict view of family?

5. Distinguish between productive and reproductive work. Which type of work is more valued?

6. How is family related to social inequality in society?

7. How has family created racial divisions and boundaries?

8. Describe at least three major changes in American family life since 1900.

9. How did the Industrial Revolution destroy the household-based economy and lead to the breadwinner system?

10. According to Kingsley Davis, what strains and demographic factors led to the collapse of the breadwinner system?

11. Describe at least three major changes in Japanese family life since 1900.

12. What caused the *ie* family system to fall? What system replaced it?

13. Explain: "Japan does not have a couple's culture."

14. What is a "parasite single"? Explain the "new single concept."

15. How is Japan's employment system connected to the country's low fertility rate?

16. In general, how do economic arrangements shape the character of sexual stratification?

17. What is *intimacy at a distance*? What factors gave rise to this phenomenon?

18. How has the status of children been affected by industrialization?

19. How do increases in life expectancy alter the composition of the family?

20. What is "caregiver burden"? Is caregiving only a burden?

21. What are some of the major differences between the elderly-caregiver relationship in Japan and the United States?

Concept Application (also in study guide)

Consider the concepts listed below. Match one or more of the concepts with each scenario. Explain your choices.

      a. Aging populations
      b. Caregiver burden
      c. Exogamy
      d. Fertility rate
      e. Reproductive work

Scenario 1

Sam works hard at his job in the factory. His supervisor knows him as a diligent, focused employee. He barely missed a day of work during his first five years on the job. A year ago, however, Sam's mother was diagnosed with Alzheimer's disease, and he decided to have her cared for in his home. But costs mounted quickly, and after a few months, Sam could only afford to have the home health care worker visit three days a week. Now, Sam is struggling to balance his job responsibilities with caring for his mother. He's almost exhausted his supply of sick days, his lack of concentration at work has caused some costly mistakes, and his supervisor's patience is at an end. Something's got to give (Guttchen and Pettigrew 2000, p. 31).

Scenario 2

Recently, the largest circulation Jewish newspaper in the country carried an opinion article pronouncing, with equanimity, that "the Jewish taboo on mixed marriage has clearly collapsed." Around the same time, and more startlingly, the *New York Times* published a photograph taken at the nuptials of a male rabbi and a female Protestant minister, a rite that was itself blessed by an assemblage of priests, ministers, and rabbis, all standing together under a Jewish wedding canopy. What this powerfully suggestive photograph tells us is not just that many American Jews, including at least some of the rabbis among them, have abandoned long-standing communal norms, but that they, again including at least some of the rabbis among them, seem to have replaced those norms with an entirely new set of beliefs about what constitutes an authentic expression of Judaism—and what, if anything, lies beyond the limits of such expression. Long in the building, the intermarriage crisis is now propelling a massive transformation of American Jewish life (Wertheimer 2001).

Scenario 3

Next year, for the first time in history, people over 60 will outnumber kids 14 or younger in industrial countries. Even more startling, the population of the Third World, while still comparatively youthful, is aging faster than that of the rest of the world. In France, for example, it took 140 years for the proportion of the population age 65 or older to double from 9 percent to 18 percent. In China, the same feat will take just 34 years; in Venezuela, it will take 22 years (Longman 1999, p. 30).

Scenario 4

Italians have stopped making babies; the nation is aging fast; and, according to the country's chief statistical body, [Italian] women now bear 1.2 babies apiece. Only the Spaniards, in Western Europe, are as unproductive. At last count, in 1996, deaths had outpaced births for four years in a row. If Italy's population is slightly up, it is thanks to the 178,000 immigrants who took up legal residence two years ago (The Economist 1998, p. 51).

Scenario 5

Sons and daughters play a crucial role in medical treatment and care for the elderly [in China]. A scarcity of medical resources, which is characteristic of developing economies, forces hospitals to rely on the work of family members to provide food, purchase and administer medicine, deliver and pick up lab tests and x-rays, and monitor and bathe the patients. Relatives draw on their personal connections to doctors and nurses to obtain treatment and hospital beds (Otis 2001, p. 471).

Practice Test: Multiple-Choice Questions

1. Japan's _____ is a major national concern.
   a. high death rate
   b. low dependency ratio
   c. low fertility rate
   d. high infant mortality rate

2. An aging population is a label attached to a situation in which
   a. the number of elderly is increasing in a society.
   b. one out of every three people is 65 and over.
   c. the youth outnumber the elderly population.
   d. the percentage of the population age 65 and older is increasing relative to other age groups.

3. Which one of the following constitutes primary kin?
   a. mother, father, sister, brother
   b. mother's mother, mother's father, sister's son
   c. brother's daughter's son
   d. brother's daughter's son's son

4. A functionalist would argue that families are structured to
   a. devalue reproductive work.
   b. maintain and perpetuate social inequalities.
   c. replace the members of society who die.
   d. foster racial divisions and boundaries.

5. Which one of the following countries has one of the <u>highest</u> fertility rates in the world?
   a. Yemen
   b. Hong Kong
   c. Japan
   d. Italy

6. Ji-wu lives in a household where his father is unemployed, but his mother works 35 hours per week at a job she has held for five years. According to the U.S. Census Bureau, Ji-wu lives in a household with
   a. a dead-beat dad.
   b. poverty-level income.
   c. insecure parental employment.
   d. secure parental employment.

7. Sociologist Kingsley Davis traced the initial rise in the divorce rate in the United States to the breadwinner system and specifically to
   a. the two-income system.
   b. increased employment opportunities for women.
   c. the shift of economic production to outside the home.
   d. women's entry into the labor market.

8. Sociologist Kingsley Davis argues that once the divorce rate reached a certain threshold, more married women seriously considered seeking employment to protect themselves in case of divorce. That threshold was _____ percent.
   a. 5
   b. 10
   c. 20
   d. 50

9. The breadwinner system is an outcome of an economic arrangement. That arrangement is
   a. socialism.
   b. agriculture-based economy.
   c. information- or service-based economy.
   d. capitalism (Industrial Revolution).

10. Under this system, "the man's economic role became, in one sense, more important to the family, for he was the link between the family and the wider market economy." *This* system is
    a. the extended family.
    b. the dual income family.
    c. the breadwinner system.
    d. traditional households.

11. Which one of the following occupying powers abolished the *ie* family system in Japan?
    a. Britain
    b. China
    c. Germany
    d. the United States

12. In Japan, the population of working single adults (22 and older) that live with their parents while contributing little to household expenses is known as
    a. the baby boomlet.
    b. spoiled singles.
    c. parasite singles.
    d. mama's boys and girls.

13. Sociologist Kaku Sechiyama argues that the key to establishing a work environment supportive of women is to
    a. establish an equal opportunity/affirmative action program.
    b. pay women *not* to have children.
    c. adopt the U.S. employment model.
    d. create a system that imposes housework, child rearing, and elder care duties on men.

14. Sociologist Randall Collins maintains that the ideology of _____ is at the heart of sexual stratification.
    a. sexual property
    b. gender polarization
    c. sexism
    d. capitalism

15. _____ are characterized by the presence of a non-householder class consisting of propertyless laborers and servants.
    a. Low-technology tribal societies
    b. Fortified households
    c. Private households
    d. Advanced market economies

16. Someone praises a Japanese mother whose son earned a grade of 100% on a math exam by saying, "He is very smart, isn't he?" Which one of the following represents her likely response?
    a. I know. He studied so hard.
    b. No. He is not so smart. He was just lucky.
    c. Yes. He is just naturally good at math.
    d. I don't know how he got to be so smart.

17. "Intimacy at a distance" is a term used to describe a situation in which norms specify that
    a. elders should not interfere in the lives of adult children.
    b. couples should practice celibacy until marriage.
    c. parents should not act as pals to their children.
    d. couples should lead separate lives.

18. _____ is the extent to which caregivers believe that their emotional balance, physical health, social life, and financial status suffer because of their caregiver role.
    a. Selective perception
    b. Self-fulfilling prophecy
    c. Life chances
    d. Caregiver burden

19. In which racial/ethnic category are men in the U.S. more likely to be caregivers to elderly family members?
    a. Asian and Pacific Islander
    b. Hispanic
    c. black
    d. white

20. In most countries, including the United States, disabled and frail elderly persons are most likely to be cared for by
    a. daughters or daughters-in-law.
    b. sons or sons-in-law.
    c. nursing home attendants.
    d. private care nurses.

True/False Questions

1. T F The family is an unchanging, stable entity.

2. T F There is no concrete group that can be universally identified as a family.

3. T F Endogamy refers to norms requiring or encouraging people to choose partners that are of different religion, race, ethnicity, or social class.

4. T F It appears that American males do more housework than their counterparts in Japan.

5. T F Viewed over a span of 100 years, the structure of the American family has changed quite dramatically.

6. T F As human muscle and time became less important to the production process, children lost their economic value.

7. T F For the most part, Japanese women are expected to quit working when they marry or have children.

8. T F In Japan, approximately 50 percent of elderly persons reside in nursing homes.

9. T F Today, divorce dissolves marriages at the same rate that death did 100 years ago.

10. T F As human muscle and time became less important to the production process, children lost their economic value.

- **U.S. Bureau of the Census**
  Families and Living Arrangements
  www.census.gov/population/www/socdemo/hh-fam.html

  The U.S. Bureau of the Census posts data and reports on a the family including demographic and other social characteristics.

- **U.S. Bureau of the Census**
  Older (55+) Population (see also Elderly (65+) Population below)
  www.census.gov/population/www/socdemo/age.html#older

  The U.S. Bureau of the Census posts data and reports on a variety of older populations including demographic and other social characteristics.

- **Japan Times Online**
  http://www.japantimes.co.jp
  For the latest news, opinions, and coverage of social and cultural events check on *Japanese Times* Online.

- **Web Japan Gateway for All Japanese Information**
  http://web-japan.org/stat/
  Search crime, education, economy, medical care, opinion surveys, leisure, politics, and other links for information on all things Japanese.

Applied Research

Put together a family tree tracing the maternal and paternal side of the family. Interview parents and grandparents. How far can each party go back in time?

Background Notes: Japan

Size: About the size of California
Capital: Tokyo (14 million)
Population (2006): 127.5 million
Ethnic groups: Ethnic groups: Japanese; Korean (0.5%)
Infant mortality rate: 2.8/1000
Life expectancy: male 78 years; female 85 years.
Work force (2005): 67 million
Workforce by Sector: *services—42%; trade, manufacturing, mining, and construction—46%; agriculture, forestry, fisheries—5%; government—3%.*

PEOPLE

Japan's population, currently some 128 million, has experienced a phenomenal growth rate during the past 100 years as a result of scientific, industrial, and sociological changes, but this has recently slowed down because of falling birth rates. In 2005, Japan's population declined for the first time, two years earlier than predicted. High sanitary and health standards produce a life expectancy exceeding that of the United States.

Japan is an urban society with only about 4 percent of the labor force engaged in agriculture. Many farmers supplement their income with part-time jobs in nearby towns and cities. About 80 million of the urban population are heavily concentrated on the Pacific shore of Honshu and in northern Kyushu. Major population centers include Metropolitan Tokyo with approximately 14 million; Yokohama with 3.3 million; Osaka with 2.6 million; Nagoya with 2.1 million; Sapporo with 1.6 million; Kyoto with 1.5 million; Kobe with 1.4 million; and Kitakyushu, Kawasaki, and Fukuoka with 1.2 million each. Japan faces the same problems that confront urban industrialized societies throughout the world: overcrowded cities, congested roads, air pollution, and rising juvenile delinquency.

Shintoism and Buddhism are Japan's two principal religions. Shintoism is founded on myths and legends emanating from the early animistic worship of natural phenomena. Since it was unconcerned with problems of afterlife, which dominate Buddhist thought, and since Buddhism easily accommodated itself to local faiths, the two religions comfortably coexisted, and Shinto shrines and Buddhist temples often became administratively linked. Today, many Japanese are adherents of both faiths.

Beyond the three traditional religions, many Japanese today are turning to a great variety of popular religious movements normally lumped together under the name "new religions." These religions draw on the concept of Shintoism, Buddhism, and folk superstition and have developed, in part, to meet the social needs of elements of the population. The officially recognized new religions number in the hundreds, and total membership is reportedly in the tens of millions.

Japan's industrialized, free market economy is the second-largest in the world. Its economy is highly efficient and competitive in areas linked to international trade, but productivity is far lower in protected areas, such as agriculture, distribution, and services. After achieving one of the highest economic growth rates in the world from the 1960s through the 1980s, the Japanese economy slowed dramatically in the early 1990s when the "bubble economy" collapsed; this was marked by plummeting stock and real estate prices.

Japan's reservoir of industrial leadership and technicians, well-educated and industrious work force, high savings and investment rates, and intensive promotion of industrial development and foreign trade produced a mature industrial economy. Japan has few natural resources, and trade helps it earn the foreign exchange needed to purchase raw materials for its economy.

Japan's long-term economic prospects are considered good, and it has largely recovered from its worst period of economic stagnation since World War II. Real GDP in Japan grew at an average of roughly 1 percent yearly in the 1990s, compared to growth in the 1980s of about 4 percent per year. The Japanese economy is now in its longest postwar expansion after more than a decade of stagnation. Real growth in 2005 was 2.7 percent and was 2.2 percent in 2006.

Only 15 percent of Japan's land is arable. The agricultural economy is highly subsidized and protected. With per hectare crop yields among the highest in the world, Japan maintains an overall agricultural self-sufficiency rate of about 40 percent on fewer than 5.6 million cultivated hectares (14 million acres). Japan normally produces a slight surplus of rice but imports, primarily from the United States, large quantities of wheat, corn, sorghum, and soybeans. Japan is the largest market for U.S. agricultural exports.

Given its heavy dependence on imported energy, Japan has aimed to diversify its sources. Since the oil shocks of the 1970s, Japan has reduced dependence on petroleum as a source of energy from more than 75 percent in 1973 to about 57 percent at present. Other important energy sources are coal, liquefied natural gas, nuclear power, and hydropower.

Deposits of gold, magnesium, and silver meet current industrial demands, but Japan is dependent on foreign sources for many of the minerals essential to modern industry. Iron ore, coke, copper, and bauxite must be imported, as must many forest products.

Japan's labor force consists of some 67 million workers, 40 percent of whom are women. Labor union membership is about 12 million.

Japan is the world's second-largest economy and a major economic power both in Asia and globally. Japan has diplomatic relations with nearly all independent nations and has been an active member of the United Nations since 1956. Japanese foreign policy has aimed to promote peace and prosperity for the Japanese people by working closely with the West and supporting the United Nations.

In recent years, the Japanese public has shown a substantially greater awareness of security issues and increasing support for the Self Defense Forces. This is in part due to the Self Defense Forces' success in disaster relief efforts at home and its participation in peacekeeping operations, such as in Cambodia in the early 1990s and Iraq in 2005-2006. However, there are still significant political and psychological constraints on strengthening Japan's security profile. Although a military role for Japan in international affairs is highly constrained by its constitution and government policy, Japanese cooperation with the United States through the 1960 U.S.-Japan Security Treaty has been important to the peace and stability of East Asia. Currently, there are domestic discussions about possible reinterpretation or revision of Article 9 of the Japanese constitution. Prime Minister Abe has made revising or reinterpreting the Japanese constitution a priority of his administration. All postwar Japanese governments have relied on a close relationship with the United States as the foundation of their foreign policy and have depended on the Mutual Security Treaty for strategic protection.

While maintaining its relationship with the United States, Japan has diversified and expanded its ties with other nations. Good relations with its neighbors continue to be of vital interest. After the signing of a peace and friendship treaty with China in 1978, ties between the two countries developed rapidly. Japan extended significant economic assistance to the Chinese in various modernization projects and supported Chinese membership in the World Trade Organization (WTO). Japan's economic assistance to China is now declining. In recent years, however, Chinese exploitation of gas fields in the East China Sea has raised Japanese concerns given the disagreement over the demarcation of their maritime boundary. Prime Minister Abe's October 2006 visits to Beijing and Seoul helped to improve strained relations with China and South Korea following Prime Minister Koizumi's visits to Yasukuni Shrine. At the same time, Japan maintains economic and cultural but not diplomatic relations with Taiwan, with which a strong bilateral trade relationship thrives.

The U.S.-Japan alliance is the cornerstone of U.S. security interests in Asia and is fundamental to regional stability and prosperity. Despite the changes in the post-Cold War strategic landscape, the U.S.-Japan alliance continues to be based on shared vital interests and values. These include stability in the Asia-Pacific region, the preservation and promotion of political and economic freedoms, support for human rights and democratic institutions, and securing of prosperity for the people of both countries and the international community as a whole.

Japan provides bases and financial and material supports to U.S. forward-deployed forces, which are essential for maintaining stability in the region. Under the U.S.-Japan Treaty of Mutual Cooperation and Security, Japan hosts a carrier battle group, the III Marine Expeditionary Force, the 5th Air Force, and elements of the Army's I Corps. The United States currently maintains approximately 50,000 troops in Japan, about half of whom are stationed in Okinawa.

Over the past decade, the alliance has been strengthened through revised Defense Guidelines, which expand Japan's noncombatant role in a regional contingency, the renewal of our agreement on Host Nation Support of U.S. forces stationed in Japan, and an ongoing process called the Defense Policy Review Initiative (DPRI). The DPRI redefines roles, missions, and capabilities of alliance forces and outlines key realignment and transformation initiatives, including reducing the number of troops stationed in Okinawa, enhancing interoperability and communication between our respective commands, and broadening our cooperation in the area of ballistic missile defense.

Implementation of these agreements will strengthen our capabilities and make our alliance more sustainable. After the tragic events of September 11, 2001, Japan has participated significantly with the global war on terrorism by providing major logistical support for U.S. and coalition forces in the Indian Ocean.

Because of the two countries' combined economic and technological impact on the world, the U.S.-Japan relationship has become global in scope. The United States and Japan cooperate on a broad range of global issues, including development assistance, combating communicable disease, such as the spread of HIV/AIDS and avian influenza, and protecting the environment and natural resources. Both countries also collaborate in science and technology in such areas as mapping the human genome, research on aging, and international space exploration. As one of Asia's most successful democracies and its largest economy, Japan contributes irreplaceable political, financial, and moral support to U.S.-Japan diplomatic efforts. The United States consults closely with Japan and the Republic of Korea on policy regarding North Korea. In Southeast Asia, U.S.-Japan cooperation is vital for stability and for political and economic reform. Outside Asia, Japanese political and financial support has substantially strengthened the U.S. position on a variety of global geopolitical problems, including the Gulf, Middle East peace efforts, and the Balkans. Japan is an indispensable partner on UN reform and the second largest contributor to the UN budget. Japan broadly supports the United States on nonproliferation and nuclear issues. The U.S. supports Japan's aspiration to become a permanent member of the United Nations Security Council.

U.S. economic policy toward Japan is aimed at increasing access to Japan's markets and two-way investment, stimulating domestic demand-led economic growth, promoting economic restructuring, improving the climate for U.S. investors, and raising the standard of living in both the United States and Japan. The U.S.-Japan bilateral economic relationship—based on enormous flows of trade, investment, and finance—is strong, mature, and increasingly interdependent. Further, it is firmly rooted in the shared interest and responsibility of the United States and Japan to promote global growth, open markets, and a vital world trading system. In addition to bilateral economic ties, the U.S. and Japan cooperate closely in multilateral forums, such as the WTO, Organization for Economic Cooperation and Development, the World Bank, and the International Monetary Fund, and regionally in the Asia-Pacific Economic Cooperation forum (APEC).

Japan is a major market for many U.S. products, including chemicals, pharmaceuticals, films and music, commercial aircraft, nonferrous metals, plastics, and medical and scientific supplies. Japan also is the largest foreign market for U.S. agricultural products, with total agricultural exports valued at $9.7 billion excluding forestry products. Revenues from Japanese tourism to the United States reached nearly $13 billion in 2005.

Answers

Concept Applications
1. Caregiver burden
2. Exogamy
3. Aging populations
4. Fertility rate
5. Reproductive work

Multiple Choice
1. c page 331
2. d page 331
3. a page 332
4. c page 335
5. a page 338
6. d page 337
7. c page 345
8. c page 345
9. d page 343
10. c page 343

11. d page 348
12. c page 349
13. d page 350
14. a page 351
15. b page 351
16. b page 353
17. a page 355
18. d page 360
19. a page 361
20. a page 362

True/False
1. F page 299
2. T page 300
3. F page 304
4. T page 307
5. T page 308
6. T page 308
7. T page 309
8. F page 316
9. T page 322
10. T page 324

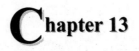Chapter 13

Education

Study Questions

1. Why was the education system of the European Union chosen as the emphasis for the Education chapter?

2. What impressions did Europeans have of early American education?

3. Distinguish between schooling, formal education, and informal education.

4. What are some of the functions schools perform to contribute to the smooth operation of society?

5. What kinds of factors do conflict theorists emphasize when they analyze systems of education?

6. What is functional illiteracy? Expand on the statement, "Illiteracy is a product of one's environment."

7. What is foreign language illiteracy? Why are Americans more likely to be illiterate in a foreign language?

8. Explain: "Only a handful of countries in the world give a significant share of their population the opportunity to go to college."

9. How does the United States compare with the European Union on providing its population opportunities to attend college?

10. What is tracking? What is the rationale for tracking? Is this rationale supported by research? Why does tracking persist?

11. How does Europe and the U.S. differ in regard to tracking?

12. Explain how the self-fulfilling prophecy can affect students' academic achievements.

13. Distinguish between formal and hidden curriculum. Give examples.

14. What is "spelling baseball"? What do children learn when they engage in such educational activities? Why?

15. Distinguish between the promise of education and the reality.

16. Explain: "Schools are the stage on which society's crises play out."

17. According to Coleman, how did the adolescent subculture emerge?

18. What are the major characteristics of the adolescent status system? How does it reflect values of the society? How does it affect education?

19. What did James Coleman uncover about American schools? What was the most controversial finding?

20. How were Coleman's findings "used"? What happened when his recommendation to bus students was implemented?

Concept Applications

Consider the concepts listed below. Match one or more of the concepts with each scenario. Explain your choices.

    a. Ability grouping
    b. Formal education
    c. Functionally illiterate
    d. Hidden curriculum
    e. Informal education
    f. Schooling
    g. Self-fulfilling prophecy
    h. Status system
    i. Tracking

Scenario 1

"Many of the deaf are functional illiterates…. Hans Furth, a psychologist whose work is concerned with the cognition of the deaf … argues that the congenitally deaf suffer from 'information deprivation.' There are a number of reasons for this. First, they are less exposed to the 'incidental' learning that takes place out of school—for example, to that buzz of conversation that is the background of ordinary life; to television, unless it is captioned, etc. Second, the content of deaf education is meager compared to that of hearing children; so much time is spent teaching deaf children speech—one must envisage between five and eight years of intensive tutoring—that there is little time for transmitting information, culture, complex skills, or anything else.

"Yet the desire to have the deaf speak, the insistence that they speak—and from the first, the odd superstitions that have always clustered around the use of sign language, to say nothing of the enormous investment in oral schools—allowed this deplorable situation to develop, practically unnoticed except by deaf people, who themselves being unnoticed had little to say in the matter." (Sacks 1989:28-29)

Scenario 2

"In 1897, Captain Richard Pratt arrived in Sioux country to enlist Sioux children for his Carlisle Indian Industrial School, the first and most famous of what would become a whole system of off-reservation boarding schools for Indian students. Eighty-four Sioux children from Pine Ridge and Rosebud, about two-thirds boys and mainly from prominent families, returned east with the stern captain. Neither parent nor pupil foresaw the short hair, the starched shirts and squeaky boots, the Christian names, or the other trappings…. Head shaving and even shackling with a ball and chain were common punishments for Indian pupils who ran away or spoke in their native tongue. Suppressing the Sioux language was high among both the Indian Bureau's educational priorities and the reasons Sioux parents kept children at home." (Lazarus 1991:101-03)

Scenario 3

"Given a paycheck and the stub that lists the usual deductions, 26 percent of adult Americans cannot determine if their paycheck is correct. Thirty-six percent, given a W-4 form, cannot enter the right numbers of exemptions in the proper places on the form. Forty-four percent, when given a series of 'help-wanted' ads, cannot match their qualifications to the job requirements. Twenty-two percent cannot address a letter well enough to guarantee that it will reach its destination. Twenty-four percent cannot add their own correct return address to the same envelope. Twenty percent cannot understand an 'equal opportunity' announcement. Over 60 percent, given a series of 'for sale' advertisements for products new and used, cannot calculate the difference between prices for a new rated and used appliance." (Kozol 1985:9)

Scenario 4

"The development of IQ tests lent an air of objectivity to the placement of procedures used to separate children for instruction....Test pioneer Lewis Terman wrote in 1916: 'At every step in the child's progress, the school should take account of his vocational possibilities. Preliminary investigations indicate that an IQ below 70 rarely permits anything better than unskilled labor; that range from 70 to 80 is pre-eminently that of semi-skilled labor, from 80 to 100 that of skilled or ordinary clerical labor, from 100 to 110 or 115 that of the semi-professional pursuits; and that above, these are the grades of intelligence which permit one to enter the professions or the larger fields of business....This information will be a great value in planning the education of a particular child and also in planning the differentiated curriculum here recommended.'" (Oakes 1985, p.36)

179

Practice Test: Multiple-Choice Question

1. Early antidotal evidence from British observers suggests that the U.S. system of education seems to value
   a. dedicated study.
   b. income generation and wealth creation.
   c. experience with other ways of life.
   d. accumulated knowledge.

2. Which one of the following characteristics applies to the process of informal education?
   a. purposeful
   b. systematic
   c. spontaneous
   d. planned

3. Early school reformers in the United States viewed education as
   a. a setting promoting diversity and multiculturalism.
   b. a liberating force.
   c. a place where students could argue about the value of ideas.
   d. the vehicle for "Americanizing" a culturally and linguistically diverse population.

4. Although all countries in the world have education-based programs that address social problems, _____ is unique in that education is viewed as the primary solution to many of its problems.
   a. the United States
   b. Japan
   c. Canada
   d. Mexico

5. Philanthropist Bill Gates argues that in the United States, there is an "acceptance of a tiering approach, where over a third of students never graduate, and another third are trapped in a situation where they don't have the skills that are going to give them a good lifetime outcome." Gates' comment corresponds to which sociological perspective?
   a. Structural strain theorists
   b. Conflict theorists
   c. Symbolic interactionists
   d. Functionalists

6. Approximately _____ percent of U.S. students that study abroad attend schools in EU countries.
   a. 20
   b. 40
   c. 50
   d. 80

7. Which racial/ethnic category of high school graduates is most likely to drop out of high school?
   a. females
   b. males
   c. Hispanic
   d. Native Americans

8. Which one of the following countries does <u>not</u> have a national curriculum that applies to all types of schools, grades, and subjects?
   a. Austria
   b. Germany
   c. the United States
   d. Portugal

9. Which one of the following responses to the question, "What is the most important thing you learned or have done so far in class?" is the one mostly likely to be made by a student placed on the college preparatory track?
   a. "I think the most important thing I have learned so far is to come into math class and get out folders."
   b. "I have learned to be more imaginative."
   c. "I have learned nothing that I'd use in later life."
   d. "To be honest, I have learned nothing."

10. A self-fulfilling prophecy begins with
    a. accurate assessment of a situation.
    b. a hidden curriculum.
    c. misguided parenting.
    d. a false definition of a situation.

11. Most European vocational school programs are equivalent in rigor to U.S. _____ programs.
    a. vocational school
    b. general studies
    c. college prep
    d. charter school

12. "Spelling baseball" and the various lessons other than spelling that students learn while playing the game represents _____ curriculum.
    a. formal
    b. hidden
    c. unintended
    d. planned

13. To be absurd means
    a. to make connections between unlike things.
    b. to fear failure.
    c. to envy success.
    d. to fear achievement.

14. The first textbooks in the United States were modeled after catechisms. Catechisms are
    a. novels that emphasize moral principles.
    b. rule books.
    c. volumes of books containing past and present knowledge arranged in alphabetical order.
    d. short books covering religious principles written in question-answer format.

15. Sociologist James Coleman defined _____ as a "small society—one that has most of its important interactions within itself, and maintains only a few threads of connection with the outside adult society."
    a. white flight
    b. schooling
    c. the adolescent subculture
    d. formal education

16. According to Coleman's research, which one of the following is a characteristic of the adolescent status system?
    a. For the most part, peer groups are less influential in students' lives than are teachers.
    b. Under no conditions was the brightest male popular.
    c. The female student identified as the brightest has the most friends.
    d. The most admired girls are cheerleaders and those girls that are successful with the boys.

17. In *Equality of Educational Opportunity* (the Coleman Report), the single most important variable for explaining differences in test scores across various ethnic groups was
    a. ethnicity.
    b. funding.
    c. family background.
    d. personal motivation.

18. Sociologist James Coleman wrote, "With families sorting themselves out residentially along economic and racial lines, and with schools tied to residence, the end result is the demise of the common school attended by children from all economic levels." Coleman was writing about the effects of
    a. busing and white flight.
    b. hidden curriculum.
    c. self-fulfilling prophecy.
    d. informal curriculum.

19. According to Coleman, students show their discontent with school by
    a. their high rate of absenteeism.
    b. getting involved with and acquiring things they can call their own.
    c. acting up in the classroom whenever teachers turn their backs.
    d. skipping out on detention.

20. Which of the following is not one of the major findings reported in "Equality of Educational Opportunity"?
    a. Variations in the quality of a school did not have much effect on student test scores.
    b. The social class of one's classmates had a significant effect on student test scores.
    c. School expenditures are an important predictor of educational attainment.
    d. Schools bring little to bear on a child's achievement independent of the child's immediate environment.

True/False Questions

1. T F   In some European Union countries, mandatory foreign language study begins as early as age five.

2. T F   In the strict sense of the word, a person who cannot read a map is illiterate.

3. T F   Every government in the world seems to think that its education system is failing in major ways.

4. T F   Throughout the U.S., public education has always been in a state of crisis and under reform.

5. T F   A national or centralized curriculum sets achievement targets but allows schools to set their own curriculum.

6. T F   Within the European Union, the spending gap between the wealthiest and poorest countries is greater than the spending gap between the wealthiest and poorest of the 50 states.

7. T F   Jules Henry uses the example of "spelling baseball" to illustrate how the hidden curriculum works.

8. T F   Self-fulfilling prophecies begin with a false definition of a situation.

9. T F   Sociologist James Coleman maintains that the adolescent society   penalizes academic achievement.

10. T F   Relative to other systems of education, the U.S. system seems to promote a math curriculum that presents the subject in a less interesting and realistic way.

Internet Resources

- **Condition of Education**
  http://nces.ed.gov/programs/coe/
  The Condition of Education is an annual report prepared by the U.S. Department of Education   National Center for Educational Statistics. The full text of the report is online, and it covers "trends in enrollments, student achievement, dropout rates, degree attainment, long-term outcomes of education, and education financing".

- **European Commission**
  http://ec.europa.eu/index_en.htm
  Te main features of the Commission's web site include "the latest official press releases, photos and live TV coverage of EU affairs, details of forthcoming events, links to the policies administered and implemented by the Commission, … and direct links to its key information services

Applied Research

Go to the What Works Clearinghouse at www.whatworks.ed.gov. The Clearinghouse was established in 2002 by the U.S. Department of Education's Institute of Education Sciences (IES) "to provide educators, policymakers, researchers, and the public with a central and trusted source of scientific evidence of what works in education." Write a paper highlighting two or three programs that work.

Size: Less than one-half the size of the U.S.
Capital: Tokyo (14 million)
Population (2006): 490.4 million
Infant mortality rate: 4.8/1000.
Life expectancy: male 75.6 years; female 82 years
Work force (2007): 227 million
Workforce by Sector: *services—*67.1%; *trade, manufacturing, mining, and
construction—*27.2%; *agriculture, forestry, fisheries—*4.3%.

27 countries: Austria, Belgium, Bulgaria, Cyprus, Czech Republic, Denmark, Estonia,
Finland, France, Germany, Greece, Hungary, Ireland, Italy, Latvia, Lithuania,
Luxembourg, Malta, Netherlands, Poland, Portugal, Romania, Slovakia, Slovenia, Spain,
Sweden, UK; note - Canary Islands (Spain), Azores and Madeira (Portugal), French
Guiana, Guadeloupe, Martinique, and Reunion (France) are sometimes listed separately
even though they are legally a part of Spain, Portugal, and France; candidate countries:
Croatia, Macedonia, Turkey

Internally, the EU is attempting to lower trade barriers, adopt a common currency, and move toward convergence of living standards. Internationally, the EU aims to bolster Europe's trade position and its political and economic power. Because of the great differences in per capita income among member states (from $7,000 to $69,000) and historic national animosities, the EU faces difficulties in devising and enforcing common policies. For example, since 2003, Germany and France have flouted the member states' treaty obligation to prevent their national budgets from running more than a three percent deficit. In 2004 and 2007, the EU admitted 10 and two countries, respectively, that are, in general, less advanced technologically and economically than the other 15. Twelve established EU member states introduced the euro as their common currency on January 1, 1999, but the UK, Sweden, and Denmark chose not to participate. Of the 12 most recent member states, only Slovenia has adopted the euro (January 1, 2007); the remaining 11 are legally required to adopt the currency upon meeting the EU's fiscal and monetary convergence criteria.

The evolution of the European Union (EU) from a regional economic agreement among six neighboring states in 1951 to today's supranational organization of 27 countries across the European continent stands as an unprecedented phenomenon in the annals of history. Dynastic unions for territorial consolidation were long the norm in Europe. On a few occasions, even country-level unions were arranged—the Polish-Lithuanian Commonwealth and the Austro-Hungarian Empire were examples—but for such a large number of nation-states to cede some of their sovereignty to an overarching entity is truly unique. Although the EU is not a federation in the strict sense, it is far more than a free-trade association, such as ASEAN, NAFTA, or Mercosur, and it has many of the attributes associated with independent nations: its own flag, anthem, founding date, and currency, as well as an incipient common foreign and security policy in its dealings with other nations. In the future, many of these nation-like characteristics are likely to be expanded. Thus, inclusion of basic intelligence on the EU has been deemed appropriate as a new, separate entity in The World Factbook. Due to the EU's special status, this description is placed after the regular country entries.

Following the two devastating World Wars of the first half of the 20th century, a number of European leaders in the late 1940s became convinced that the only way to establish a lasting peace was to unite the two chief belligerent nations—France and Germany—both economically and politically. In 1950, the French Foreign Minister Robert Schuman proposed an eventual union of all Europe, the first step of which would be the integration of the coal and steel industries of Western Europe. The following year, the European Coal and Steel Community (ECSC) was set up when six members— Belgium, France, West Germany, Italy, Luxembourg, and the Netherlands—signed the Treaty of Paris. The ECSC was so successful that within a few years, the decision was made to integrate other parts of the countries' economies. In 1957, the Treaties of Rome created the European Economic Community (EEC) and the European Atomic Energy Community (Euratom), and the six member states undertook to eliminate trade barriers among themselves by forming a common market. In 1967, the institutions of all three communities were formally merged into the European Community (EC), creating a single Commission, a single Council of Ministers, and the European Parliament. Members of the European Parliament were initially selected by national parliaments; in 1979, the first direct elections were undertaken, and they have been held every five years since. In 1973, the first enlargement of the EC took place with the addition of Denmark, Ireland, and the United Kingdom. The 1980s saw further membership expansion with Greece joining in 1981 and Spain and Portugal in 1986. The 1992 Treaty of Maastricht laid the basis for further forms of cooperation in foreign and defense policy, in judicial and internal affairs, and in the creation of an economic and monetary union, including a common currency. This further integration created the European Union (EU). In 1995, Austria, Finland, and Sweden joined the EU, raising the membership total to 15. A new currency, the euro, was launched in world money markets on January 1, 1999; it became the unit of exchange for all of the EU states except the United Kingdom, Sweden, and Denmark. In 2002, citizens of the 12 euro-area countries began using the euro banknotes and coins. Ten new countries joined the EU in 2004—Cyprus, the Czech Republic, Estonia, Hungary, Latvia, Lithuania, Malta, Poland, Slovakia, and Slovenia—and in 2007, Bulgaria and Romania joined, bringing the current membership to 27. In order to ensure that the EU can continue to function efficiently with an expanded membership, the Treaty of Nice (in force as of February 1, 2003) set forth rules streamlining the size and procedures of EU institutions. An effort to establish an EU constitution, begun in October 2004, failed to attain unanimous ratification. A new effort, undertaken in June 2007, calls for the creation of an Intergovernmental Conference to form a political agreement, known as the Reform Treaty, which is to serve as a constitution. Unlike the constitution, however, the Reform Treaty would amend existing treaties, rather than replace them.

Source: Excerpted from *World Factbook*, U.S. Bureau of Central Intelligence (2007) www.cia.gov/library/publications/the-world-factbook/geos/ee.html

Answers

Concept Applications

1. Self-fulfilling prophey
2. Formal Education, Schooling
3. Functionally illiterate
4. Ability grouping, Tracking

Multiple Choice				**True/False**		
1.	b	page 367	11.	c	page 381	1.	T	page 367
2.	c	page 368	12.	b	page 384	2.	T	page 373
3.	d	page 369	13.	a	page 384	3.	T	page 388
4.	a	page 370	14.	d	page 386	4.	T	page 388
5.	b	page 372	15.	c	page 389	5.	F	page 379
6.	c	page 375	16.	d	page 391	6.	T	page 372
7.	d	page 377	17.	c	page 392	7.	T	page 382
8.	c	page 379	18.	a	page 394	8.	T	page 381
9.	b	page 380	19.	b	page 391	9.	T	page 390
10.	d	page 381	20.	c	page 393	10.	T	page 385

Chapter 14

Religion

Study Questions

1. Why was Afghanistan chosen as the country to emphasize with regard to religion?

2. When sociologists study religion, what do they study?

3. According to Durkheim, how should sociologists approach the study of religion?

4. According to Durkheim, what are three fundamental and indispensable features of religion? How do these features figure into a definition of religion?

5. Distinguish between the sacred and the profane. What are the three major types of religion, as categorized in terms of sacred phenomena?

6. According to Durkheim, what are rituals? What are the most important outcomes of rituals?

7. Distinguish between ecclesiae, denominations, sects, established sects, and cults.

8. What are some problems with Durkheim's definition of religion? Give examples. Are there better definitions?

9. What is civil religion? What role did civil religion play in the Cold War?

10. How did Muslims come to be partners to the U.S. during the Cold War?

11. Is the question "What is religion" only of interest to sociologists? Explain.

12. What function does religion serve for the individual and the group?

13. Explain what Durkheim means by the statement, "The something out there that people worship is actually society." How is it that society is worthy of such worship?

14. Is religion strictly an integrative force? Why or why not?

15. How did Karl Marx conceptualize religion?

16. What are some criticisms of Marx's views of religion?

17. According to Weber, what role did the Protestant ethic play in the origins and development of modern capitalism? In what ways has Weber been misinterpreted?

18. What is secularization? Distinguish between Muslim views and American-European views about the causes of secularization.

19. What is fundamentalism? How are fundamentalism and secularization related?

20. What are the factors behind the surge of fundamentalism in Muslim countries?

21. Distinguish between religious and political *jihad* (including militant Islam).

22. How many militant Islamic political *jihadists* exist in the world today?

Concept Applications

Consider the concepts listed below. Match one or more of the concepts with each scenario. Explain your choices.

    a. Church
    b. Civil religion
    c. Liberation theology
    d. Mystical religions
    e. Rituals
    f. Sect

Scenario 1

"As for my own religious practice, I try to live my life pursuing what I call the Bodhisattva ideal.... The Bodhisattva idea is thus the aspiration to practice infinite compassion with infinite wisdom. As a means of helping myself in the quest, I choose to be a Buddhist monk. There are 253 rules of Tibetan monasticism (364 for nuns) and by observing them as closely as I can, I free myself from many of the distractions and worries of life. Some of these rules mainly deal with etiquette, such as the physical distance a monk should walk behind the abbot of his monastery; others are concerned with behavior. The four root vows concern simple prohibitions: namely that a monk must not kill, steal, or lie about his spiritual attainment. He must also be celibate. If he breaks any one of these, he is no longer a monk." (Gyatso 1990:204-05)

Scenario 2

"There were usually three services each day: morning, midafternoon, and evening. A ram's horn summoned everyone to the nine-o'clock morning service, at which time people would leave their camps and congregate in the shed. Sunday was the biggest day of the week, and for many years it was also the day of the Lovefeast. Bread and water were passed around and people would make their testimonials. During the evening service there would inevitably be an altar call, often accompanied by a lot of shouting." (Jenkins 1996:562)

Scenario 3

"The 'miracle' was Brazil's accelerated economic growth between 1968 and 1975; Brazil moved from twenty-first to fourteenth in rank among developing countries, based upon per capita GNP. The 'miracle' did not help most Brazilians, however. The imbalances in the distribution of wealth were made yet worse. The Brazilian bishops have openly denounced the 'Brazilian miracle' for the poverty it has engendered. They have attacked the economic policies that have pushed thousands of peasant farmers off the lands their families have farmed for generations, and they have questioned development projects (such as the exploitation of the Amazon) which displaced the native Indians and poor farmers but brought them no benefit. Indeed, one observer has concluded that 'the church has become the primary institutional focus of dissidence in the country.'" (McGuire 1987:215)

Scenario 4

"Mennonites trace their roots to a small group of Christians after 1530 who sought a reformation even more radical than those advocated by Lutherans and Calvinists. They were called Mennonites after Menno Simons, one of their early leaders. Their most distinctive practice is adult baptism offered only to those who have made a decision to follow Christ's teachings." (Lorimer 1989:212)

Scenario 5

"The old fascist marching songs were sung, a moment of silence was observed for all who died defending the fatherland, and the gathering was reminded that today was the 57th anniversary of the founding of Croatia's Nazi-allied wartime government. Then came the most chilling words of the afternoon.

'For Home!' shouted Anto Dapic, surrounded by bodyguards in black suits and crew cuts.

'Ready!' responded the crowd of 500 supporters, their arms rising in a stiff Nazi salute.

The call and response—the Croatian equivalent of 'Sieg!' 'Heil!'—was the wartime greeting used by supporters of the fascist Independent Sate of Croatia that governed the country for most the Second World War and murdered hundreds of thousands of Jews, Serbs and Croatian resistance fighters." (Hedges 1997:3Y)

Practice Test: Multiple-Choice Questions

1. _____ wrote "To define 'religion,' to say what it is, is not possible at the start of a presentation such as this. Definition can be attempted, if at all, only at the conclusion of the study."
   a. Karl Marx
   b. Max Weber
   c. Emile Durkheim
   d. George H. Mead

2. Sacred things can include books, buildings, days, and places. From a sociological point of view sacredness stems from
   a. the item itself.
   b. an item's symbolic power.
   c. the meaning assigned to it by God.
   d. the bible.

3. Which of the following statements does <u>not</u> apply to Native Spirituality?
   a. There are probably as many native religions as there are Indian tribes.
   b. The basic tenets of Native Spirituality can be found in the "Great Book".
   c. None of the native religions have man-made churches in the Judeo-Christian sense.
   d. Religious beliefs are tied to nature.

4. Some of the most well-known _____ religions include Judaism, Confucianism, Christianity, and Islam.
    a. sacramental
    b. prophetic
    c. mystical
    d. profane

5. Confession, immersion, and fasting are examples of
    a. mystical acts.
    b. ecclesiae.
    c. rituals.
    d. sacraments.

6. _____ include(s) everything that is not sacred.
    a. Powerful symbols
    b. Evil
    c. The profane
    d. Exorcism

7. Durkheim used the word "church" to designate a group whose members do all but which one of the following?
    a. hold the same beliefs with regard to the sacred and the profane
    b. behave in the same way in the presence of the sacred
    c. gather together to affirm commitment to beliefs and practices
    d. adhere to the belief that the religion members follow is one of many true religions

8. In Islam the most pronounced split occurred after the death of Prophet Muhammad over the issue of Muhammed's successor. That split is between
    a. Sunni and Shia.
    b. Hezbollah and Druze.
    c. Iranian Sunni and Iraqi Shia.
    d. Muslims and Jews.

9. In light of Durkheim's definition of religion, which one of the following does not qualify as a religious phenomenon?
    a. displays of patriotism
    b. 21-gun salutes
    c. national holidays
    d. traffic jams in which everyone gets out of their cars to interact

10. When the Soviet Union invaded Afghanistan in 1979, the U.S.
    a. let the Soviets control the country.
    b. dropped an atomic bomb.
    c. sent 250,000 troops to the region.
    d. supported the Afghan freedom fighters, known as the *mujahidin*.

11. The Soviet Union invaded Afghanistan in
    a. 1979.
    b. 1994.
    c. 1945.
    d. 1890.

12. The United States worked with _____ to recruit 35,000 Muslims from 43 countries to fight with their Afghan brothers against the Soviet Union.
    a. Iran
    b. Pakistan
    c. Tajikistan
    d. India

13. Functionalists maintain that religion must serve some vital social function because
    a. there are very few atheists in the world.
    b. all people turn to religion in times of deep distress.
    c. some form of religion has existed as long as humans have been around.
    d. people who communicate with their god find extraordinary strength.

14. Which one of the following sociologists observed that whenever a group of people have a strong conviction, that conviction almost always takes on a religious character?
    a. Robert Coles
    b. Robert K. Merton
    c. Emile Durkheim
    d. Max Weber

15. If religion were truly an integrative force
    a. there would be no conflict or tensions among religious groups within the same society.
    b. everyone would have the same religion.
    c. there would be fewer struggles between the political and the religious.
    d. everyone would be a member of a religion.

16. Durkheim observed that whenever a group of people has a strong conviction
    a. religious values are secondary to the conviction.
    b. that conviction always takes on a religious character.
    c. they fight among themselves.
    d. they work to make the world a better place.

17. _____ is/are an example of a religion that emerged in the United States in the 1930s as a vehicle of protest or change.
    a. Liberation theology
    b. The Quakers
    c. Black Shia
    d. Nation of Islam

18. Danielle believes that God has foreordained all things including the salvation or damnation of individual souls. This belief is known as
    a. liberation theology.
    b. secularization.
    c. predestination.
    d. fundamentalism.

19. Weber maintained that the Protestant ethic
    a. caused capitalism to come into being.
    b. led to the rise of fundamentalism.
    c. was a significant force in the emergence of capitalism.
    d. must be present in a society if it is to achieve economic success.

20. Daniel Pipes estimates that there are _____ million persons "who do not accept the particulars" of militant Islam but are sympathetic and supportive of the anti-American stance.
    a. 1
    b. 10
    c. 100
    d. 500

True/False Questions

1. T F In studying religions, sociologists must assume that there are no religions that are false.

2. T F Rituals can be codes of conduct aimed at governing the performance of everyday activities such as eating.

3. T F All the major religions encompass splinter groups that have sought to preserve the integrity of their religion.

4. T F Cults often dissolve after their leader dies.

5. T F Durkheim maintained that whenever a group of people has a strong conviction, it almost always takes on a religious character.

6. T F Archeological evidence suggests that Jesus was at least 6 feet 2 inches tall.

7. T F Karl Marx maintained that people need the comfort of religion in order to make the world bearable and justify their existence in it.

8. T F The Protestant Ethic <u>caused</u> capitalism to emerge.

9. T F Fundamentalism is a process by which religious influences on thought and behavior are reduced.

10. T F There are no features common to all religions

Internet Resources

- **About Specific Religions, Faith Groups, Ethical Systems, etc.**
  http://www.religioustolerance.org/var_rel.htm
  This website have information on the at least 40 major religions and is supported by the Ontario Consultants on Religious Tolerance whose aims are to promote tolerance of minority religions, offer useful information on controversial religious topics, and expose hatred and misinformation about any religion.

- **Afghanistan News.net**
  http://afghanistannews.net/
  For the latest news on Afghanistan visit this website.

- **Afghanistan's Website**
  http://www.afghanistans.com/
  Learn about Afghanistan's land and resources, the people, the climate, past and present flags, proverbs, alphabet, music, view photos and daily news coverage.

Applied Research

Visit a religious service hosted by a religious organization with which you are unfamiliar. Use sociological concepts from Chapter 14 to frame an analysis of that experience.

Size: slightly smaller than Texas
Capital: Kabul
Population (2005): 31.1 million
Ethnic groups: Pashtun, Tajik, Hazara, Uzbek, Turkmen, Aimaq, Baluch, Nuristani, Kizilbash
Religions: Sunni Muslim 80%, Shi'a Muslim 19%, other 1%
Infant mortality rate: 165/1,000 births
Life expectancy at birth: female 42.27 years; male 42.66 years

Afghanistan's ethnically and linguistically mixed population reflects its location astride historic trade and invasion routes leading from Central Asia into South and Southwest Asia. While population data is somewhat unreliable for Afghanistan, Pashtuns comprise the largest ethnic group at 38 to 44 percent of the population, followed by Tajiks (25%), Hazaras (10%), Uzbeks (6-8%), Aimaqs, Turkmen, Baluchs, and other small groups. Dari (Afghan Farsi) and Pashto are official languages. Dari is spoken as a first language by more than one-third of the population and serves as a lingua franca for most Afghans, though Pashto is spoken throughout the Pashtun areas of eastern and southern Afghanistan. Tajik and Turkic languages are spoken widely in the north. Smaller groups throughout the country speak more than 70 other languages and numerous dialects.

Afghanistan is an Islamic country. An estimated 80 percent of the population is Sunni, following the Hanafi school of jurisprudence; the remainder of the population—primarily the Hazara ethnic group—is predominantly Shi'a. Despite attempts to secularize Afghan society during the years of communist rule, Islamic practices pervade all aspects of life. In fact, Islam served as a principal basis for expressing opposition to communism and the Soviet invasion. Islamic religious tradition and codes, together with traditional tribal and ethnic practices, have an important role in personal conduct and dispute settlement. Afghan society is largely based on kinship groups, which follow traditional customs and religious practices. Kinship groups are somewhat less common in urban areas.

During the 19th century, collision between the expanding British Empire in the subcontinent and czarist Russia significantly influenced Afghanistan in what was termed "The Great Game." British concern over Russian advances in Central Asia and growing Russian influence in Persia culminated in two Anglo-Afghan wars. The first (1839-42) resulted in the destruction of a British army and is remembered today as an example of the ferocity of Afghan resistance to foreign rule. The second Anglo-Afghan war (1878-80) was sparked by Amir Sher Ali's refusal to accept a British mission in Kabul. This conflict brought Amir Abdur Rahman to the Afghan throne. During his reign (1880-1901), the British and Russians officially established through the demarcation of the Durand Line the boundaries of what would become modern Afghanistan. The British retained effective control over Kabul's foreign affairs.

Afghanistan remained neutral during World War I, despite German encouragement of anti-British feelings and Afghan rebellion along the borders of British India. The Afghan king's policy of neutrality was not universally popular within the country, however.

The Taliban had risen to power in the mid 90s in reaction to the anarchy and warlordism that arose after the withdrawal of Soviet forces. Many Taliban had been educated in madrassas in Pakistan and were largely from rural southern Pashtun backgrounds. In 1994, the Taliban developed enough strength to capture from a local warlord the city of Kandahar and proceeded to expand its control throughout Afghanistan, occupying Kabul in September 1996. By the end of 1998, the Taliban occupied about 90 percent of the country, limiting the opposition largely to a small mostly Tajik corner in the northeast and the Panjshir valley.

The Taliban sought to impose an extreme interpretation of Islam—based upon the rural Pashtun tribal code—on the entire country and committed massive human rights violations, particularly directed against women and girls. The Taliban also committed serious atrocities against minority populations, particularly the Shi'a Hazara ethnic group and killed noncombatants in several well-documented instances. In 2001, as part of a drive against relics of Afghanistan's pre-Islamic past, the Taliban destroyed two Buddha statues carved into cliff faces outside the city of Bamiyan.

From the mid-1990s, the Taliban provided sanctuary to Osama bin Laden, a Saudi national who had fought with the mujahideen resistance against the Soviets. The Taliban provided a base for his and other terrorist organizations. Bin Laden provided both financial and political support to the Taliban. Bin Laden and his Al-Qaida group were charged with the bombing of the U.S. Embassies in Nairobi and Dar Es Salaam in 1998, and in August 1998, the United States launched a cruise missile attack against bin Laden's terrorist camp in southeastern Afghanistan. Bin Laden and Al-Qaida have acknowledged their responsibility for the September 11, 2001, terrorist attacks against the United States.

Following the Taliban's repeated refusal to expel bin Laden and his group and to end its support for international terrorism, the U.S. and its partners in the anti-terrorist coalition began a military campaign on October 7, 2001, targeting terrorist facilities and various Taliban military and political assets within Afghanistan. Under pressure from U.S. military and anti-Taliban forces, the Taliban disintegrated rapidly, and Kabul fell on November 13, 2001.

Afghan factions opposed to the Taliban met at a United Nations-sponsored conference in Bonn, Germany, in December 2001 and agreed to restore stability and governance to Afghanistan—creating an interim government and establishing a process to move toward a permanent government. Under the "Bonn Agreement," an Afghan Interim Authority was formed and took office in Kabul on December 22, 2001, with Hamid Karzai as Chairman. The Interim Authority held power for approximately six months while preparing for a nationwide "Loya Jirga" (Grand Council) in mid-June 2002 that decided on the structure of a Transitional Authority. The Transitional Authority, headed by President Hamid Karzai, renamed the government as the Transitional Islamic State of Afghanistan (TISA). One of the TISA's primary achievements was the drafting of a constitution that was ratified by a Constitutional Loya Jirga on January 4, 2004.

On October 9, 2004, Afghanistan held its first national democratic presidential election. More than eight million Afghans voted, 41 percent of whom were women. Hamid Karzai was announced as the official winner on November 3 and was inaugurated on December 7 for a five-year term as Afghanistan's first democratically elected president. On December 23, 2004, President Karzai announced new cabinet appointments, naming three women as ministers.

An election was held on September 18, 2005, for the "Wolesi Jirga" (lower house) of Afghanistan's new bicameral National Assembly and for the country's 34 provincial councils. Turnout for the election was about 53 percent of the 12.5 million registered voters. The Afghan constitution provides for indirect election of the National Assembly's "Meshrano Jirga" (upper house) by the provincial councils and by reserved presidential appointments. The first democratically elected National Assembly since 1969 was inaugurated on December 19, 2005. Younus Qanooni and Sigbatullah Mojadeddi were elected Speaker of the Wolesi Jirga and Meshrano Jirga, respectively.

The government's authority is growing, although its ability to deliver necessary social services remains largely dependent on funds from the international donor community. Between 2001 and 2006, the United States committed over $12 billion to the reconstruction of Afghanistan. At an international donors' conference in Berlin in April 2004, donors pledged a total of $8.2 billion for Afghan reconstruction over the three-year period of 2004 to 2007. At the end of January 2006, the international community gathered in London and renewed its political and reconstruction support for Afghanistan in the form of the Afghanistan Compact.

With international community support, including more than 40 countries participating in Operation Enduring Freedom and NATO-led International Security Assistance Force (ISAF), the government's capacity to secure Afghanistan's borders in order to maintain internal order is increasing. Responsibility for security for all of Afghanistan was transferred to ISAF in October 2006. As of November 2006, some 40,000 Afghan National Army (ANA) soldiers had been trained, along with some 60,000 police, including border and highway police.

The main source of income in the country is agriculture, and during its good years, Afghanistan produces enough food and food products to provide for the people, as well as to create a surplus for export. The major food crops produced are corn, rice, barley, wheat, vegetables, fruits, and nuts. In Afghanistan, industry is also based on agriculture and pastoral raw materials. The major industrial crops are cotton, tobacco, madder, castor beans, and sugar beets. The Afghan economy continues to be overwhelmingly agricultural, despite the fact that only 12 percent of its total land area is arable and less than 6 percent is currently cultivated. Agricultural production is constrained by an almost total dependence on erratic winter snows and spring rains for water; irrigation is primitive. Relatively few people use machines, chemical fertilizers, or pesticides.

Overall, agricultural production declined dramatically following severe drought, as well as sustained fighting, instability in rural areas, and deteriorated infrastructure. During 2003, the easing of the drought and the end of civil war produced the largest wheat harvest in 25 years. Wheat production was an estimated 58 percent higher than in 2002. However, the country still needed to import an estimated one million tons of wheat to meet its requirements for the 2003 year. Millions of Afghans, particularly in rural areas, remained dependent on food aid.

Opium has become a source of cash for many Afghans, especially following the breakdown in central authority after the Soviet withdrawal. Opium-derived revenues probably constituted a major source of income for the two main factions during the civil war in the 1990s. Opium is easy to cultivate and transport and offers a quick source of income for impoverished Afghans. Afghanistan produced a record opium poppy crop in 2006, supplying 91 percent of the world's opium. Much of Afghanistan's opium production is refined into heroin and is either consumed by a growing regional addict population or exported, primarily to Western Europe.

Afghanistan has begun counter-narcotics programs, including the promotion of alternative livelihoods, public information campaigns, targeted eradication policies, interdiction of drug shipments, as well as law enforcement and justice reform programs. These programs were first implemented in late 2005. In June 2006, the United Nations Office on Drugs and Crime estimated that the Afghan Government eradicated over 15,000 hectares of opium poppy.

Afghanistan is endowed with natural resources, including extensive deposits of natural gas, petroleum, coal, copper, chromite, talc, barites, sulfur, lead, zinc, iron ore, salt, and precious and semiprecious stones. Unfortunately, ongoing instability in certain areas of the country, remote and rugged terrain, and an inadequate infrastructure and transportation network have made mining these resources difficult, and there have been few serious attempts to further explore or exploit them.

The most important resource has been natural gas, which was first tapped in 1967. At their peak during the 1980s, natural gas sales accounted for $300 million a year in export revenues (56 percent of the total). Ninety percent of these exports went to the Soviet Union to pay for imports and debts. However, during the withdrawal of Soviet troops in 1989, Afghanistan's natural gas fields were capped to prevent sabotage by the mujahidin. Restoration of gas production has been hampered by internal strife and the disruption of traditional trading relationships following the collapse of the Soviet Union. Trade in smuggled goods into Pakistan once constituted a major source of revenue for Afghan regimes, including the Taliban, and it still figures as an important element in the Afghan economy, although efforts are underway to formalize this trade.

Since the fall of the Taliban in 2001, many nations have assisted in a great variety of humanitarian and development projects all across Afghanistan. The United Nations, World Bank, Asian Development Bank, and other international agencies have also given aid. Schools, clinics, water systems, agriculture, sanitation, government buildings, and roads are being repaired or built.

Afghanistan is one of the most heavily mined countries in the world; mine-related injuries number up to 100 per month, and an estimated 200,000 Afghans have been disabled by landmine/unexploded ordinances (UXO) accidents. As of March 2005, the United Nations Mine Action Program for Afghanistan had approximately 8,000 Afghan personnel, 700 demobilized soldiers, 22 international staff, and several NGOs deployed in Afghanistan. The goal of the program is to remove the impact of mines from all high-impact areas by 2007 and to make Afghanistan mine-free by 2012. Between January 2003 and March 2005, a total of 2,354,244 mines and pieces of UXOs were destroyed. Training programs are also being used to educate the public about the threat and dangers of land mines. The number of mine victims was reduced from approximately 150 a month in 2002 to less than 100 a month in 2004.

Afghanistan has had the largest refugee repatriation in the world in the last 30 years. The return of refugees is guided by the Ministry of Refugees and Repatriation (MORR) and supported by the United Nations High Commissioner for Refugees (UNHCR), the International Organization of Migration (IOM), the United Nations Children's Fund (UNICEF), the World Food Program (WFP), the World Health Organization (WHO), and a number of other national and international NGOs. As of December 2006, approximately three million Afghans remained in neighboring countries. The U.S. provided more than $350 million to support Afghan refugees, returnees, and other conflict victims between September 2001 and March 2006.

Excerpted from U.S. CIA (2007), Country Background Notes
http://www.state.gov/r/pa/ei/bgn/5380.htm

Answers

1. Mystical religions, Rituals
2. Church, Rituals
3. Liberation theology
4. Sect
5. Civil religion

Multiple Choice

1.	b	page 402
2.	b	page 403
3.	b	page 405
4.	b	page 405
5.	c	page 407
6.	c	page 406
7.	d	page 408
8.	a	page 410
9.	d	page 411
10.	d	page 412

11.	a	page 412
12.	b	page 412
13.	c	page 417
14.	c	page 418
15.	a	page 418
16.	b	page 418
17.	d	page 422
18.	c	page 423
19.	c	page 423
20.	d	page 427

True/False

1.	T	page 402
2.	T	page 407
3.	T	page 410
4.	T	page 410
5.	T	page 418
6.	F	page 419
7.	T	page 420
8.	F	page 423
9.	F	page 425
10.	F	page 428

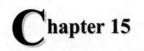hapter 15

Population and Urbanization

Study Questions

1. Why is India the focus of a chapter on population? How do China and the U.S. compare to India in terms of population size?

2. Distinguish between crude birth rate, age-sex specific birth rate, and total fertility rate.

3. How is population size determined? Explain.

4. Define crude death rate and infant mortality rate.

5. Why is India's mortality rate lower than the rate of the United States?

6. What are the various types of migration and immigration? What are push and pull factors?

7. What is doubling time? At what point in history did the world's population reach 1 billion? How long did it take to reach 2 billion? 3 billion? 6 billion?

8. How are labor intensive economies different from core economies?

9. Why is Stage I of the demographic transition called the stage of "high potential growth"?

10. According to the model of the demographic transition, which factors contributed to a decline in the death rate? To a rise and then an eventual decline in fertility?

11. Why does the demographic transition model not apply to India and other labor intensive countries?

12. What factors contribute to declines in total fertility? To what extent has India realized these factors?

13. What is a population pyramid? What shapes can it take?

14. When referring to countries, how is the dichotomy "industrialized—industrializing" misleading? What are more appropriate terms?

15. What is a demographic trap?

16. What is urbanization?

17. What is a mega city?

18. How does urbanization in labor-intensive poor countries differ from urbanization in core economies?

19. Distinguish among a central city, a suburb, and a nonmetropolitan area.

Concept Application (also in study guide)

Consider the concepts listed below. Match one or more of the concepts with each scenario. Explain your choices.

    a. Cohort
    b. Demographic trap
    c. Internal migration
    d. Migration
    e. Positive checks
    f. Pull factors
    g. Push factors
    h. Stationary pyramids
    i. Urbanization

Scenario 1

    "By 2025, over 1 billion people in Africa and southern Asia will live under conditions of water scarcity. Many North African and Middle Eastern countries are already faced with absolute water scarcity. In Jordan and Israel, over 3,000 people compete for every flow unit of renewable water. By 2025, virtually all North African countries will be faced with high levels of population pressure on their scarce water resources. And, except for Turkey, all of Western Asia will also experience the highest levels of water scarcity." (Falkenmark and Widstrand 1992:20)

Scenario 2

    "The reality is of course that, since World War II, tens of millions of people have opted to leave the quiet of the countryside, either 'expelled' by drought, disease, or political strife or drawn by dreams broadcast over transistor radios. Some, like the half-million Guatemalan Indians who travel each winter with their wives and families to the Pacific lowlands to pick coffee and cotton or to cut sugarcane, do so in order to survive in their villages during the rest of the year. But for most, migration is a one-way experience because those who break with their families and communities, their traditional language, clothes, and food change too much to be able to return." (Riding 1986:8)

Scenario 3

      The population pyramid for Denmark looks more like a rectangle than a pyramid. "Each cohort is about the same size as every other one because the birth rate and the death rate have been low and relatively constant for a long time. This means that each age group is about the same size at birth, and since relatively few people die before old age, the cohorts remain close in size until late in life when mortality rates must rise and eat away at the top of the rectangle." (McFalls 1991:22-23)

Scenario 4

      "The villages were as quiet as death…. In one village, I remember we had as our guide a tall, middle-aged peasant who had blue eyes and a straw-colored beard. When he spoke of the famine in all those villages hereabouts, he struck his breast and tears came into his eyes. He led us into timbered houses where Russian families were hibernating and waiting for death. In some of them, they had no food of any kind. There was one family I saw who left an indelible mark on my mind. The father and mother were lying on the floor when we entered and were almost too weak to rise. Some young children were on a bed above the stove, dying of hunger. A boy of eighteen lay back in a wooden settle against the window sill in a kind of coma. These people had nothing to eat—nothing at all." (Gibbs 1987:494)

Scenario 5

      "The theme of this book is the lives and reactions of certain patients in a unique situation—and the implications which these hold out for medicine and science. These patients are among the few survivors of the great sleeping-sickness epidemic fifty years ago, and their reactions are those brought about by a remarkable new 'awakening' drug (L-Dopa). The lives and responses of these patients, which have no real precedent in the entire history of medicine, are presented in the form of extended case histories or biographies." (Sacks 1989:1)

Practice Test: Multiple-Choice Questions

1. India's population reached 1 billion in May of 2000. The country might have reached this milestone in 1989 had it not been
    a. for the devastating natural disasters of the prior 20 years.
    b. the first country in the world to adopt a national family planning program.
    c. the first country in the world to close its doors to immigrants.
    d. for a history of emigration.

2. The United States accounts for _____ percent of the world's population.
    a. 30
    b. 10.5
    c. 4.6
    d. 1.9

3. Use the following information to calculate the age-specific birth rate for India; total births in year—27,116,788; number of women ages 15-54—300,527,000. The age-specific birth rate is _____ per 1,000.
    a. 51.99
    b. 5199
    c. 76.5
    d. 90.2

4. A subspecialty within sociology that focuses on the study of human population is
    a. epidemiology.
    b. ethnomethodology.
    c. demography.
    d. conflict theory.

5. A population's age and sex composition is commonly depicted as a
    a. three-dimensional graph.
    b. cohort.
    c. population pyramid.
    d. demographic transition.

6. Country Y has a population of 149.3 million people. Life expectancy is 73 years for men and 79 years for women. The total fertility is below replacement level. The population pyramid for this country would be
    a. expansive.
    b. constrictive.
    c. stationary.
    d. triangular.

7. Stewart is moving out of his hometown because there are no jobs. The reason he is moving is called a
   a. push factor.
   b. pull factor.
   c. demographic.
   d. self-motivating factor.

8. Which one of the following factors represents an example of a <u>pull</u> factor?
   a. unfavorable climate
   b. employment opportunities
   c. discrimination
   d. natural disaster

9. Within the United States, the greatest amount of internal migration is movement
   a. within the same county.
   b. from one county to another.
   c. from one state to another.
   d. into adjacent counties.

10. Almost 30 percent of India's total population changed residences in the past 10 years. This movement is dominated by short-distance, rural-to-rural movements within India. Sociologists classify this kind of migration as
    a. immigration.
    b. emigration.
    c. internal migration.
    d. international.

11. The demographic transition
    a. is a two-stage model of population growth.
    b. depicts the history of birth and death rates in labor-intensive poor countries.
    c. depicts the history of disease in core economies.
    d. depicts the history of population growth in Western Europe and North America.

12. 50/1000 is believed to be the highest _____ rate possible for any society.
    a. death
    b. fertility
    c. marriage
    d. birth

13. In demographic terms, the Black Death is an example of
    a. a mortality crisis.
    b. a life expectancy crisis.
    c. a tragedy.
    d. a degenerative disease.

14. According to Thomas Malthus, epidemics, war, and famine are examples of
    a. positive checks.
    b. demographic traps.
    c. demographic gaps.
    d. catastrophic events.

15. The <u>least</u> important reason for the decline in death rates in Western societies is
    a. improvement in agricultural technology.
    b. improvement in sanitation.
    c. medical advances.
    d. proper disposal of sewage.

16. Which one of the following countries is least likely to be in Stage III of the demographic transition?
    a. The United States
    b. Germany
    c. India
    d. Japan

17. Urbanization includes all but which one of the following characteristics?
    a. increase in the number of cities
    b. growth of the population living in cities
    c. rural-to-urban migration
    d. urban-to-rural migration

18. The _____ is considered "the most illustrious and most flourishing commercial association that ever existed in any age or country."
    a. Bank of India
    b. General Electric Company
    c. East India Company
    d. Domino Sugar Company

19. India was once a colony of
    a. the United States.
    b. Portugal.
    c. Spain.
    d. Britain.

20. Which one of the following U.S. cities is <u>not</u> among the world's top 30 economies?
    a. San Francisco
    b. Boston
    c. Houston
    d. Philadelphia

True/False Question

1. T F China is the country with the largest population in the world.

2. T F Japan is among the 10 most populous countries in the world.

3. T F The death rate in India and the United States is about the same.

4. T F Historically, emigration rates for India were especially low during times of major famine and epidemics.

5. T F Mexico is the country that attracts the greatest number of all Americans who live abroad.

6. T F India's sex ratio is skewed in favor of males.

7. T F The Industrial Revolution was an event confined to the world's core economies.

8. T F In India, female sterilization appears to be a major form of contraception.

9. T F In India, total fertility has increased over the past four decades.

10. T F Humans produce enough food each year to feed the world's population.

Internet Resources

Population Reference Bureau
- **http://www.prb.org**
  The Population Reference Bureau covers births, deaths, migration, and other population topics as they relate to the United States and other countries. Examples of the hundreds of articles on the webpage include "Why Do Canadians Outlive Americans?," "The Lives and Times of the Baby Boomers," and "Hurricanes, Population Trends, and Environmental Change."

- **The *Times of India***
  http://timesofindia.indiatimes.com
  The *Times of India* covers major news and current events relevant to India and India's relationship to the world.

- **Census of India**
  http://www.censusindia.net/
  Census data, maps and vital statistics about India can be found on this site from the Office of the Registrar General, India.

Applied Research

The text points out that Hurricane Katrina "pushed" as estimated 1 million people out of their homes to nearby and far away locations. Access the excel spread sheet from the U.S. Census Bureau identifying affected coastal counties and their population sizes pre- and post Katrina for Alabama, Louisiana, Mississippi, and Texas (www.census.gov/Press-Release/www/emergencies/gulfcoast_impact_estimates.xls). Determine the most affected county in each state and the most affected state. To determine the most affected, figure population change by subtracting 2005 population estimate from 2006 estimate and then dividing by the 2005 estimate. The largest negative percentages represent the most affected areas.

Background Notes: India

Size: About one-third the size of the U.S.
Capital: New Delhi (12.8 million)
Population (2006): 107.4 million
Ethnic groups: Indian-Spanish (mestizo) 60%; Indian 30%; Caucasian 9%; other 1%
Religions: Roman Catholic 89%; Protestant 6%; other 5%
Infant mortality rate: 54.6/1,000
Life expectancy: 64.7 years
Work force: 450 million
Workforce by Sector: Agriculture—62%; industry and commerce—22%; services and government—12%; transport and communications—4%

Although India occupies only 2.4 percent of the world's land area, it supports over 15 percent of the world's population. Only China has a larger population. Almost 33 percent of Indians are younger than 15 years of age. About 70 percent live in more than 550,000 villages and the remainder in more than 200 towns and cities. Religion, caste, and language are major determinants of social and political organization in India today. The government has recognized 18 official languages. Hindi, the national language, is the most widely spoken, although English is a national lingua franca. Although 82 percent of its people are Hindu, India also is the home of more than 138 million Muslims—one of the world's largest Muslim populations. The population also includes Christians, Sikhs, Jains, Buddhists, and Parsis.

The Hindu caste system reflects Indian occupational and socially defined hierarchies. Ancient Sanskrit sources divide society into four major categories: priests (Brahmin), warriors (Kshatriya), traders (Vaishya), and farmers/laborers (Shudra). Although these categories are understood throughout India, they describe reality only in the most general terms. They omit, for example, the tribes and those once known as "untouchables." In reality, Indian society is divided into thousands of jatis—local, endogamous groups based on occupation—and is organized hierarchically according to complex ideas of purity and pollution. Despite economic modernization, laws countering discrimination against the lower end of the caste structure, and outlawing "untouchability," the caste system remains an important source of social identification and a potent factor in the political life of the country. Nevertheless, the government has made strong efforts to minimize the importance of caste through active affirmative action and social policies. Moreover, caste has been diluted, if not subsumed, in the economically prosperous and heterogeneous cities where an increasing percentage of India's population lives. In the countryside, expanding education, land reform, and economic opportunity through access to information, communication, transport, and credit have lessened the harshest elements of the caste system.

India's population is estimated at nearly 1.1 billion and is growing at 1.3 percent a year. It has the world's 12th largest economy—and the third largest in Asia behind Japan and China—with total GDP of around $797 billion. Services, industry, and agriculture account for 51 percent, 28 percent, and 21 percent of GDP, respectively. Nearly two-thirds of the population depends on agriculture for its livelihood. About 28 percent of the population lives below the poverty line, but there is a large and growing middle class of 325 to 350 million with disposable income for consumer goods.

India is continuing to move forward with market-oriented economic reforms that began in 1991. Recent reforms include liberalized foreign investment and exchange regimes, industrial decontrol, significant reductions in tariffs and other trade barriers, reform and modernization of the financial sector, significant adjustments in government monetary and fiscal policies, and safeguarding intellectual property rights.

Real GDP growth for the fiscal year ending March 31, 2006, was 8.4 percent, up from 7.7 percent growth in the previous year. Growth for the year ending March 31, 2007, is expected to be between 7.8 and 8.3 percent. Foreign portfolio and direct investment inflows have risen significantly in recent years. They have contributed to the $166 billion in foreign exchange reserves by mid-September 2006. Government receipts from privatization were about $3 billion in fiscal year 2003-04.

However, economic growth is constrained by inadequate infrastructure, a cumbersome bureaucracy, corruption, labor market rigidities, regulatory and foreign investment controls, the "reservation" of key products for small-scale industries, and high fiscal deficits. The outlook for further trade liberalization is mixed. India eliminated quotas on 1,420 consumer imports in 2002 and has announced its intention to continue to lower customs duties. However, the tax structure is complex, with compounding effects of various taxes.

The United States is India's largest trading partner. Bilateral trade in 2005 was $26.8 billion. Principal U.S. exports are diagnostic or lab reagents, aircraft and parts, advanced machinery, cotton, fertilizers, ferrous waste/scrap metal, and computer hardware. Major U.S. imports from India include textiles and ready-made garments, Internet-enabled services, agricultural and related products, gems and jewelry, leather products, and chemicals.

The rapidly growing software sector is boosting service exports and modernizing India's economy. Revenues from the information technology (IT) industry reached a turnover of $23.6 billion in 2005-06. Software exports crossed $22 billion in FY2005-06. IT and business process outsourcing (BPO) exports are projected to grow at nearly 27 to 30 percent during 2006-07. Personal computer penetration is 14 per 1,000 persons. The cellular/mobile market surged to 140 million subscribers by November 2006. The country has 54 million cable TV customers.

The United States is India's largest investment partner, with a 13 percent share. India's total inflow of U.S. direct investment is estimated at more than $5 billion through 2005-06. Proposals for direct foreign investment are considered by the Foreign Investment Promotion Board, and they generally receive government approval. Automatic approvals are available for investments involving up to 100 percent foreign equity, depending on the kind of industry. Foreign investment is particularly sought after in power generation, telecommunications, ports, roads, petroleum exploration/processing, and mining.

India's external debt was $125 billion in 2005-06, up from $123 billion in 2004-05. Foreign assistance was approximately $3.8 billion in 2005-06, with the United States providing about $126 million in development assistance. The World Bank plans to double aid to India to almost $3 billion a year, with focus on infrastructure, education, health, and rural livelihoods.

Source: Excepted from U.S. Department of State: Background Notes
http://www.state.gov/r/pa/ei/bgn/3454.htm

Answers

Concept Applications
1. Demographic trap
2. Internal migration, Pull factors, Push factors, Urbanization
3. Stationary pyramids
4. Positive checks
5. Cohort

Multiple Choice

1.	b	page 433	11.	d	page 442
2.	c	page 434	12.	d	page 444
3.	c	page 435	13.	a	page 444
4.	c	page 434	14.	a	page 445
5.	c	page 441	15.	c	page 446
6.	b	page 441	16.	c	page 447
7.	a	page 436	17.	d	page 446
8.	b	page 436	18.	c	page 448
9.	d	page 439	19.	d	page 448
10.	c	page 440	20.	a	page 454

True/False

1.	T	page 443
2.	T	page 434
3.	T	page 436
4.	F	page 437
5.	T	page 439
6.	T	page 442
7.	F	page 447
8.	T	page 449
9.	F	page 449
10.	T	page 451

hapter 16

Social Change

Study Questions

1. Why is Greenland the focus of the chapter on social change?

2. What is social change? Why are sociologists interested in tipping points?

3. What questions do sociologists ask when they study social change?

4. What has changed since 1750? Why is 1750 an important date?

5. What about industrialization and mechanization has contributed to fossil fuel dependence?

6. Distinguish between global interdependence and globalization. How are they connected to fossil fuel dependence?

7. What is rationalization and value-rational thought? How have the two contributed to fossil fuel dependence?

8. What is McDonaldization? How has it contributed to fossil fuel dependence?

9. What is urbanization? How has it contributed to fossil fuel dependence?

10. What is the information explosion? What technological innovations are responsible for this phenomenon?

11. What factors does Orrin Klapp identify as the causes underlying distorted and exaggerated presentation of information?

12. When thinking about social change, why is it difficult to pinpoint a single cause of change?

13. What is an innovation? Distinguish between basic and improving innovations. What makes an innovation sociologically significant?

14. What is the cultural base? How is the rate of change tied to the size of the cultural base?

15. What is cultural lag? Why did Ogburn emphasize the material component of culture in this theory of cultural lag?

16. Is Ogburn a technological determinist? Why or why not?

17. Ogburn maintains that one of the most urgent challenges facing people today is adapting to material innovations. Does the work of Leslie White lend support to Ogburn's thesis? Why or why not?

18. How does Kuhn define a paradigm?

19. According to Thomas Kuhn, is science simply an evolutionary process? Why or why not? Under what conditions are paradigms threatened? When does a scientific revolution occur?

20. How is conflict both a cause and an effect of social change?

21. Describe the essential dynamics of the Cold War and how those dynamics are connected to the development of the Internet.

22. From a world-system perspective, how has capitalism come to dominate the global network of economic relationships?

23. What is a social movement? What conditions are necessary for social movements to occur?

24. What are the different types of social movements? Give a brief description of each.

25. Distinguish between objective deprivation and relative deprivation. How are these concepts related to social movements?

26. What are the three stages in the life of a social movement?

27. What kinds of social interactions give insights into climate change's effect on Greenland?

28. How do sociologists use the three theoretical perspectives to frame a discussion about Greenland and climate change?

29. How is the culture of Greenland's Inuit and of other Arctic peoples changing because of climate change?

30. How do ingroup and outgroup memberships related to climate change shape identity?

31. What social forces bring Greenlanders into interaction with outsiders and shape the relationship between the two groups?

32. Because of climate change, what new formal organizations have emerged in Greenland?

33. How do ideas about what constitutes deviance relate to outsiders' interest or lack of interest in Greenland?

34. How is climate change shaping life chances in Greenland and elsewhere?

35. What is the sex composition of Greenland? How might it be affected by climate change?

36. How did the U.S. military-industrial complex pull Greenland into the international arena?

37. How might climate change affect Greenland's fertility rate?

38. What are formal and informal ways outsiders are coming to learn about Greenland, other Arctic cultures, and climate change?

39. What religions did outsiders bring to Greenland?

40. What is the population size of Greenland, and is the population increasing or decreasing as a result of climate change?

41. In light of the information explosion, how does one identify credible sources about climate change?

Concept Application (also in study guide)

Consider the concepts listed below. Match one of more of the concepts with each scenario. Explain your choices.

    a. Globalization
    b. Paradigms
    c. Planned obsolescence
    d. Scientific revolution
    e. Technological determinism

Scenario 1

"It is difficult to recapture the medical world of 1800; it was a world of thought structured around assumptions so fundamental that they were only occasionally articulated as such, yet assumptions alien to a twentieth-century medical understanding…. The body was seen as a system of intake and outgo, a system that had to remain in balance if the individual [was] to remain healthy…. Equilibrium was synonymous with health, disequilibrium with illness…. The physician's most effective weapon was his ability to 'regulate the secretions' to extract blood, to promote the perspiration, the urination, or defecation that attested to his having helped the body regain its customary equilibrium." (Rosenberg 1987:71-72)

Scenario 2

"Of course, Federal Express is our largest business unit by far. It is quite simply, the largest global express transportation network ever assembled. On our first night of operations— back in April of 1973—we delivered just 186 packages to 25 U.S. cities, using a fleet of 14 small Falcon jets. Twenty-five years later, FedEx delivers about 3 million shipments every business day to 211 countries that generate better than 90 percent of the world's GDP. The Federal Express workforce has grown to about 142,000 employees…. FedEx has the largest commercial cargo fleet in the world, with 615 aircraft and about 100 more on order. That ranks us as the fourth largest airline worldwide—not just in the cargo industry but among passenger airlines as well. In addition, the FedEx ground network includes about 42,000 trucks and vans, which are linked back into our data network to provide real-time information, from pick-up to delivery. And by utilizing one of the largest interactive computer networks in the world, better than 60 percent of all FedEx transactions are now handled electronically." (Smith 1998)

Scenario 3

"In public discussions of biotechnology today, the idea of improving the human race by artificial means is widely condemned. The idea is repugnant because it conjures up visions of Nazi doctors sterilizing Jews and killing defective children. There are many good reasons for condemning enforced sterilization and euthanasia. But the artificial improvement of human beings will come, one way or another, whether we like it or not, as soon as the progress of biological understanding makes it possible. When people are offered technical means to improve themselves and their children, no matter what they conceive improvement to mean, the offer will be accepted. Improvement may mean better health, longer life, a more cheerful disposition, a stronger heart, a smarter brain, the ability to earn more money as a rock star or baseball player or business executive. The technology of improvement may be hindered or delayed by regulation, but it cannot be permanently suppressed. Human improvement, like abortion today, will be officially disapproved, legally discouraged, or forbidden, but [it will be] widely practiced. It will be seen by millions of citizens as a liberation from past constraints and injustices. Their freedom to choose cannot be permanently denied." (Dyson 1997:49)

Scenario 4

"Thomas Kuhn's seminal work *The Structure of Scientific Revolutions* affected working scientists as deeply as it moved those scholars who scrutinize what we do. Before Kuhn, most scientists followed the place-a-stone-in-the-bright-temple-of-knowledge tradition and would have told you that they hoped, above all, to lay many of the bricks, perhaps even the keystone, of truth's temple, the addictive or meliorists model of scientific progress. Now, most scientists of vision hope to forment revolution." (Gould 1987: 27)

Scenario 5

"In the 1930s, an enterprising engineer working for General Electric proposed increasing sales of flashlight lamps by increasing their efficiency and shortening their life. Instead of lasting through three batteries, he suggested that each lamp last only as long as one battery. In 1934, speakers at the Society of Automotive Engineers meetings proposed limiting the life of automobiles. These examples and others are cited in Vance Packard's classic book *The Waste Makers*." (Beder 1998)

Practice Test: Multiple-Choice Questions

1. When studying a social change, sociologists ask,
    a. "Is social change good for society?"
    b. "How can we stop social change?"
    c. "Is social change necessary?"
    d. "What are consequences of change in terms of social life?"

2. A 2007 UN report announced that climate change can no longer be denied or doubted and that human activity since _____ has "very likely" caused the rise in the planet's temperatures.
    a. 1500
    b. 1750
    c. 1850
    d. 1950

3. _____ is any significant alteration, modification, or transformation in the organization and operation of social life.
    a. Globalization
    b. Scientific revolution
    c. Social change
    d. Global interdependence

4. The most critical factor that drove the Industrial Revolution was _____.
    a. mechanization
    b. the information explosion
    c. planned obsolescence
    d. human muscle

5. _____ involves producing goods that are disposable after a single use, have a shorter life cycle than the industry is capable of producing, or go out of style quickly even though the goods can still serve their purpose.
    a. Mechanization
    b. Rationalization
    c. Planned obsolescence
    d. A tipping point

6. A company maintains, "We deliver within 30 minutes!" That company is applying which of the following McDonaldization principles?
    a. efficiency
    b. quantification/calculation
    c. predictability
    d. control

7. Which analogy did sociologist Orrin Klapp use to describe the dilemma of sorting through and keeping up with the massive amount of information being generated?
   a. A sociologist drowning in quicksand
   b. A student trying to take notes while 10 professors talk at one time
   c. A researcher working on a gigantic jigsaw puzzle while additional pieces are flowing onto the table from a funnel overhead
   d. A person entering a crowded six-lane highway with thousands of signs

8. "We invent the automobile to get us between two points faster, and suddenly we find we have to build new roads. And that means we have to invent traffic regulations...and then we have to invent a whole new organization called the highway patrol." This assessment supports the idea that
   a. necessity is the mother of invention.
   b. if a new invention is to come into being, the cultural base must be large enough to support it.
   c. invention is the mother of necessity.
   d. if people have the power to create material innovations, they also have the power to destroy them.

9. _____ are situations in which the same invention is created at about the same time by two or more people working independently of one another.
   a. Cultural diffusions
   b. Scientific revolutions
   c. Improving inventions
   d. Simultaneous-independent inventions

10. When a new paradigm causes converts to see the world in an entirely new light and causes them to wonder how they could possibly have taken the old paradigm seriously, _____ has occurred.
    a. a scientific revolution
    b. innovation
    c. cultural lag
    d. adaptive reasoning

11. Perhaps the most outstanding feature of the Internet is that it was designed to operate
    a. from a central command station in Washington.
    b. on solar power.
    c. automatically.
    d. without a central control over it.

12. Marx believed that _____ was the first economic system capable of maximizing the immense productive potential of human labor and ingenuity.
    a. the capitalist system
    b. socialism
    c. communism
    d. a centrally planned economy

13. A social movement depends on three conditions. Which one of the following is <u>not</u> one of those conditions?
    a. an actual or imagined condition that enough people find objectionable
    b. a shared belief that something needs to be done about this condition
    c. some organized effort aimed at attracting supporters, articulating the problem, and defining a strategy
    d. enough financial support to get the movement off the ground

14. Ralf Dahrendorf wrote, "It is immeasurably difficult to trace the path on which a person…encounters other people just like himself, and at a certain point says, 'Let us join hands, friends, so that no one will push us off one by one.'" Dahrendorf was writing about
    a. globalization-from-above.
    b. social movements.
    c. globalization-from-below.
    d. the Chernobyl meltdown.

15. A_____ of climate change is a growing interest in Greenland, the Arctic, and Antarctica such that popular films are set in or give prominent attention to these locations.
    a. manifest function
    b. latent function
    c. manifest dysfunction
    d. latent dysfunction

16. Sociologists use the term _____ to describe a group with which people identify and to which they feel closely attached, particularly when that attachment is founded on opposition to another group.
    a. primary group
    b. ingroup
    c. outgroup
    d. secondary group

17. Sociologists look to identify the resources of Greenland that pulled its people into the global division of labor. From the 16th through the late 19th centuries, that resource was
    a. oil extracted from the ground.
    b. oil extracted from whales.
    c. rubber.
    d. ivory.

18. From a global perspective, people who live in _____ have the lowest access to sustainable water, with the equivalent of 2,640 gallons available to each person each year.
    a. Kuwait
    b. the United States
    c. Canada
    d. Greenland

19. The Distance Early Warning line, a radar and satellite system in place to warn of an impending ballistic missile attack against North America,
    a. stretches from Alaska through northern Canada to Greenland.
    b. stretches through Russia and into Greenland.
    c. is located exclusively in Greenland.
    d. is located along the U.S.-Canada border.

20. Signers of the Petition Web project, a group that believes climate change is part of natural weather patterns rather than increased fossil fuel use, would likely be associated with which one of the following?
    a. Conoco Phillips
    b. Netherlands Environmental Assessment Agency
    c. Canadian Centre for Climate Modeling and Analysis
    d. National Center for Atmospheric Research

True/False Questions

1. T F Since 1900, humans have burned fossil fuels to transport, among other things, people and goods.

2. T F Conflict is a key trigger of social change.

3. T F Geometric expansion can be represented by the following sequence: 1, 2, 4, 8, 16, 32...

4. T F From a sociological point of view, invention is the mother of necessity.

5. T F Some inventions, such as the bicycle, generate no conflict in society.

6. T F In a capitalist system, profit is the most important measure of success.

7. T F Capitalist responses to economic stagnation and downturn helped to create a global network of economic relationships.

8. T F Research on social movements shows that the most objectively disadvantaged people join social movements to change their condition.

9. T F From a conflict perspective, corporations and their customers will benefit from the effects of climate change on Greenland at the expense of Greenland's native peoples.

10. T F Greenland was once a colony of Denmark.

Internet Resources

- **Intergovernmental Panel on Climate Change (IPCC)**
  http://www.ipcc.ch/about/about.htm
  The full text of IPCC Special Reports on climate change can be found at this website.

- **Explore Greenland**
  http://www.greenland.com/content/english/tourist
  The website offers the official national guide to Greenland. Topics covered include "towns and places," "traveling to Greenland," "dog sledding," "wild life safari," and "icebergs and the ice cap."

Applied Research

Use Google News Search engine to follow news stories on global warming/climate change. Follow news on this topic for least a week and then write a three to four pages analysis of the major themes.

Size: slightly more than three times the size of Texas
Capital: Nuuk
Population (2006): 56,344
Ethnic groups: Greenlander 88% (Inuit and Greenland-born whites); Danish and others 12%
Infant mortality rate: 14.98/1000
Life expectancy: male 66.6 years; female 73.9 years
Work force: 32,120

Greenland, the world's largest island, is about 81 percent ice-capped. Vikings reached the island from Iceland in the 10th century. Danish colonization began in the 18th century, and Greenland was made an integral part of Denmark in 1953. It joined the European Community (now the EU) with Denmark in 1973 but withdrew in 1985 over a dispute centered on stringent fishing quotas. Greenland was granted self-government in 1979 by the Danish parliament; the law went into effect the following year. Denmark continues to exercise control of Greenland's foreign affairs in consultation with Greenland's Home Rule Government.

The economy remains critically dependent on exports of fish and substantial support from the Danish Government, which supplies about half of government revenues. The public sector, including publicly-owned enterprises and the municipalities, plays the dominant role in the economy. Several interesting hydrocarbon and mineral exploration activities are ongoing. Press reports in early 2007 indicated that two international aluminum companies were considering building smelters in Greenland to take advantage of local hydropower potential. Tourism is the only sector offering any near-term potential, and even this is limited due to a short season and high costs. Air Greenland announced plans to begin summer season direct flights to the U.S. east coast in May 2007, potentially opening a major new tourism market.

Source: Excepted from U.S. Department of Central Intelligence, World Factbook (2007), www.cia.gov/library/publications/the-world-factbook/geos/gl.html

Answers

Concept Applications
1. Paradigms
2. Globalization
3. Technological Determinism
4. Paradigms, Scientific Revolution
5. Planned Obsolescence

Multiple Choice

1. d page 459
2. b page 459
3. c page 460
4. a page 461
5. c page 464
6. b page 464
7. c page 466
8. c page 469
9. d page 469
10. a page 470

11. d page 471
12. a page 472
13. d page 474
14. b page 474
15. b page 476
16. b page 477
17. b page 478
18. a page 480
19. a page 482
20. a page 485

True/False

1. F page 459
2. T page 467
3. T page 468
4. T page 469
5. F page 471
6. T page 472
7. T page 472
8. F page 474
9. T page 476
10. T page 480